Proceedings of the Munich Symposium on Lightweight Design 2020

Simon Pfingstl · Alexander Horoschenkoff ·
Philipp Höfer · Markus Zimmermann
Editors

Proceedings of the Munich Symposium on Lightweight Design 2020

Tagungsband zum Münchner Leichtbauseminar 2020

Springer Vieweg

Editors
Simon Pfingstl
Laboratory for Product Development and Lightweight Design
Technical University of Munich
Garching, Germany

Alexander Horoschenkoff
Department of Mechanical, Automotive and Aeronautical Engineering
Munich University of Applied Sciences
Munich, Germany

Philipp Höfer
Institute of Lightweight Engineering
Universität der Bundeswehr München
Neubiberg, Germany

Markus Zimmermann
Laboratory for Product Development and Lightweight Design
Technical University of Munich
Garching, Germany

ISBN 978-3-662-63142-3 ISBN 978-3-662-63143-0 (eBook)
https://doi.org/10.1007/978-3-662-63143-0

This Springer Vieweg imprint is published by the registered company Springer-Verlag GmbH, DE part of Springer Nature.
The registered company address is: Heidelberger Platz 3, 14197 Berlin, Germany

Preface

Dear reader,

the Munich area is an important center of lightweight design: it hosts major industries, including aerospace or automotive, several renowned universities and research centers. It has been an epicenter of innovation and academic research for decades. In 2003, the Technical University of Munich, the Universität der Bundeswehr, and the University of Applied Sciences in Munich initiated the first symposium on lightweight design to promote the local networks of industry and academia. Over the years, this has turned into a national platform for the exchange of ideas and research activities. Experienced and young professionals from both, industry and academia, present their recent work at the Munich Lightweight Symposium. Summaries of most talks are now available for the first time as Proceedings (Tagungsband zum Münchner Leichtbauseminar 2020) as an online book by Springer. We are pleased to present our conference now to a larger audience and hope you enjoy it.

Best regards

January 2021

Simon Pfingstl
Alexander Horoschenkoff
Philipp Höfer
Markus Zimmermann

Organization

Program Chairs

Simon Pfingstl	Technical University of Munich, Department of Mechanical Engineering, Laboratory for Product Development and Lightweight Design
Alexander Horoschenkoff	University of Applied Sciences Munich, Department of Mechanical, Automotive and Aeronautical Engineering
Philipp Höfer	Universität der Bundeswehr München, Department of Aerospace Engineering, Institute of Lightweight Engineering
Markus Zimmermann	Technical University of Munich, Department of Mechanical Engineering, Laboratory for Product Development and Lightweight Design

Contents

Optimization

Composite Materials

Design of a motorcycle triple clamp optimised for stiffness and damping

Tobias Ehlers[1[0000-0003-4811-2186]] and Roland Lachmayer[1[0000-0002-3181-6323]]

[1] Institute of Product Development (IPeG), Gottfried Wilhelm Leibniz Universität Hannover, An der Universität 1, 30823 Garbsen, Germany
ehlers@ipeg.uni-hannover.de

Abstract. Engineers have always faced the challenge of solving conflicting objectives such as high stiffness combined with high damping. Structurally optimised components are used, especially by pushing lightweight construction. This design adaptation of component mass and stiffness generally has a negative effect on the dynamic component properties, as both the natural frequencies are shifted and component damping is reduced. In the majority of applications, the resulting vibrations are undesirable and must be reduced by suitable mechanisms. For example, vibrations in the vehicle can lead to a reduction in driving comfort or to a reduced service life.

One approach to solving conflicting objectives is the targeted integration of effects into components through additive manufacturing. In this paper, the effect-engineering on a laser beam melted motorcycle triple clamp is illustrated. The triple clamp is a highly dynamically loaded structural component where unwanted vibrations occur due to road unevenness, leading to critical hand-arm vibrations. This paper focuses on the simulative design of the triple clamp. The triple clamp is topology-optimised and extended by the effect of particle damping, so that the component is optimised in terms of stiffness, damping and mass. The optimisation also makes it possible to achieve a high degree of functional integration by saving 20 components. The effect of particle damping is experimentally evaluated by preliminary studies, which show that component damping can be increased by up to a factor of 20. The laser powder bed fusion (LPBF) makes it possible to store unmelted powder in the interior of the component in a targeted manner and thus produce particle-damped structures inside the triple clamp.

Keywords: Laser powder bed fusion (LPBF), Particle damping, Additive manufacturing (AM), Functional integration, Design for additive manufacturing (DfAM), Effect-engineering

S. Pfingstl et al. (eds.), *Proceedings of the Munich Symposium on Lightweight Design 2020*, https://doi.org/10.1007/978-3-662-63143-0_1

1 Introduction

In recent years, the design for additive manufacturing (DFAM) has been supplemented and improved by more and more methods to design and manufacture even more efficient products [1–6]. In the field of DFAM, the design of assemblies offers great potential [7,8]. Any optimisation of existing individual parts cannot be as good as re-thinking the entire assembly design [7,8]. One obstacle in the design of assemblies is the restriction of the design space due to design limitations. These design limitations can occur at points where two or more areas with conflicting design specifications meet, for example high stiffness and high damping [9]. Thus, the question arises how these limits can be overcome and shifted in order to develop more efficient products and to raise potentials.

The laser powder bed fusion (LPBF) has so far been characterised by its ability to produce stiffness-optimised components with a minimum of material. Recently, it has also become possible with this production method to integrate damping elements locally in vibration-prone areas [10–21]. For this purpose, unmelted powder is left inside the component. In keeping with the "Complexity for free" [22] principle, "Damping for free" applies to additively manufactured components, since neither mass nor costs increase with increasing damping compared to the additively manufactured reference component [21]. Here, a reduction in vibration by a factor of more than 20 can be achieved from a few Hz up to several kHz. Thus, the LPBF makes it possible to solve the conflict of objectives between high stiffness and high damping.

In this paper a method for the design of effect optimised components with regard to stiffness and damping is presented. The effect of particle damping is evaluated on laser beam melted bending beams made of AlSi10Mg. These results are then transferred to a motorcycle triple clamp as a demonstrator. Here the assembly of the triple clamp is considered in order to realise a high degree of functional integration. The motorcycle triple clamp is designed and optimised for the conflicting objectives of stiffness, mass and damping. The stiffness of the triple clamp is one of the criteria for riding safety [23]. At the same time, critical hand-arm vibrations must be reduced so as not to impair driving comfort and health [24]. The damping elements previously used in triple clamps reduce stiffness and thus driving safety. The triple clamp is simulatively optimised with regard to stiffness by carrying out a topology optimisation. Subsequently, particle-filled cavities are integrated into the component, thus integrating the effect of damping. The effect of the particle damping is estimated for the motorcycle triple clamp from the damping curves of the parameter study.

2 Theoretical background

2.1 Hand-arm vibrations

Excessive vibration in the hand-arm area can not only reduce comfort, but can also lead to illnesses over a longer period of time, which are also known as Hand-arm vibration syndrome (HAVS) [24–27]. HAVS can lead to vascular, neurological and muscular-skeletal disorders [24–26]. These disorders are recognised throughout Europe as occupational diseases [26].

Vascular disorders are characterised by pale or white fingers, especially in cold weather [24,26,28]. This disorder, known as Raynaud's phenomenon, is due to a temporary interruption of the blood supply to the fingers [24,26,28]. The sense of touch can be impaired or completely lost [24,26]. In addition, finger mobility is restricted. In general, the risk of accidents when handling machines is increased [26]. The Raynaudian phenomenon occurs in northern countries with colder climates in about 80-100 % of the workers exposed to high exposure levels at work [28]. Even in the general male population, this value is between 1.5 and 14 % [28]. The effects of the vascular disturbances depend on the amplitude, frequency, direction, impact and duration of the vibration exposure [26].

If exposure to vibration occurs over a long period of time, vascular symptoms may worsen and neurological disorders such as carpal tunnel syndrome (CTS) may also occur. [26]. The effects on healing are lower in neurological diseases than in vascular disorders [28].

Prolonged vibration stress can also lead to muscular diseases such as pain in the hand-arm area and a reduction in muscle/gripping strength and to arthritis or tendinitis [24,26]. Fig. 1 schematically shows the occurring consequences of HAV.

ISO 5349-1 can be used for the calculation and assessment of vibration exposure [26]. According to EU Directive 2002/44/EC, measures must be taken to reduce vibrations for workers from a daily exposure limit value of 2.5 m/s^2, related to 8 hours. The maximum permissible daily exposure action value is 5 m/s^2. Persons exposed to an acceleration of 2 m/s^2 for less than 8 hours can be expected to suffer minor hand-arm disorders. [26].

One of the first studies on hand-arm vibrations of motorbikes was conducted in 1997 among police motorcyclists [24,29]. It was found that finger numbness, finger stiffness and shoulder pain are common [24,29]. In addition, in everyday driving situations from approx. 2 hours of driving time, it is to be expected that the exposure to vibrations is higher than specified in international guidelines [30–32]. Further measures to reduce vibrations must therefore be considered. Vibrations on the motorcycle can be caused by the engine, by unfavourable mass moments of inertia, imbalances or by external forces such as road unevenness [23,32,33].

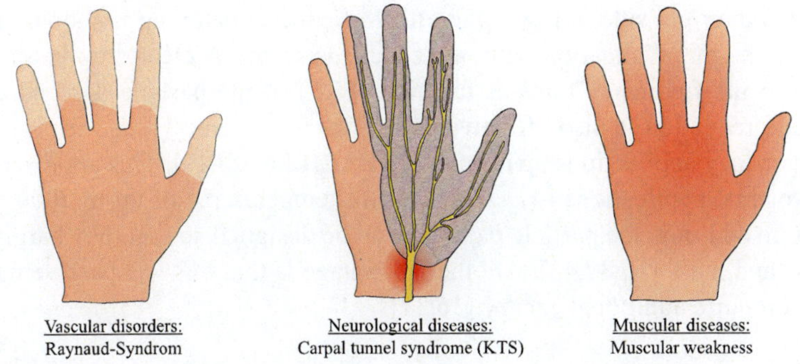

Fig. 1. Possible diseases caused by hand-arm vibrations (HAV)

2.2 Measures to reduce hand-arm vibrations on motorcycles

In the following, measures are presented for motorcycles to reduce vibrations caused by road unevenness and engine vibrations.

In order to reduce vibrations caused by uneven road surfaces, tyres play a central role due to their suspension and damping properties [23]. In addition, spring and damper elements are integrated in the stem so that the wheel can follow the uneven road profile in a vertical direction, but the motorcycle and rider do not make any vertical movements [23]. In order to optimally adjust the damping properties in the stem, semi-active or in rare cases also active damping systems are used [23]. As a rule, you can choose between different driving modes, which can be set to either comfort (high damping), sport (high driving safety combined with low damping) or a compromise of both [23]. Thus, there is always a conflict of objectives between driving comfort and driving safety. [23]

Rubber elements for engine mounting in the frame or an elastic decoupling of the handlebars from the frame help to reduce the transmission of engine vibrations to the hands [23]. It is important to note that the connection between the engine and the frame must be as rigid as possible in order to achieve sufficient driving safety. Thus there is a conflict of aims between stiffness and damping [23]. Particularly in the case of sports motorcycles, vibrations with the associated loss of comfort are accepted in order to achieve a high level of driving safety [23].

This raises the question of how high damping can be achieved at the same time as high stiffness in motorcycles, while keeping the mass low. One solution is, for example, the integration of particle damping in the handlebars [34].

2.3 Use of laser beam melted particle dampers

Laser Powder Bed Fusion (LPBF) is a manufacturing process in which the component is built up layer by layer by local melting of powder [2,35]. Furthermore, with LPBF it is possible to leave unmelted powder inside the component and thus integrate the particle damping effect into the component [16,19]. Through the combined forces of inelastic impacts and friction, either particle/particle or particle/wall interaction, kinetic energy is absorbed from the vibrating main structure resulting in increased damping [36,37]. A simplified diagram of a particle damper is shown on the left side in Fig. 2. On the right side in Fig. 2 the effect of particle damping is shown qualitatively related to the initial situation without particle damping. A clear drop in amplitude can be seen in the frequency transfer function (FRF) of the particle damped curve. This amplitude reduction is also effective over a wide frequency range from a few Hz to high-frequency applications up to several kHz [21,36,38–40]. This applies both to cold and cryogenic applications [41] and to warm temperatures of up to 2000 °C or more [37,42]. In addition, the particle dampers can be designed so that they hardly affect the stiffness and mass [10,37,43]. Another advantage is that this is a passive measure that does not require additional energy [36,37].

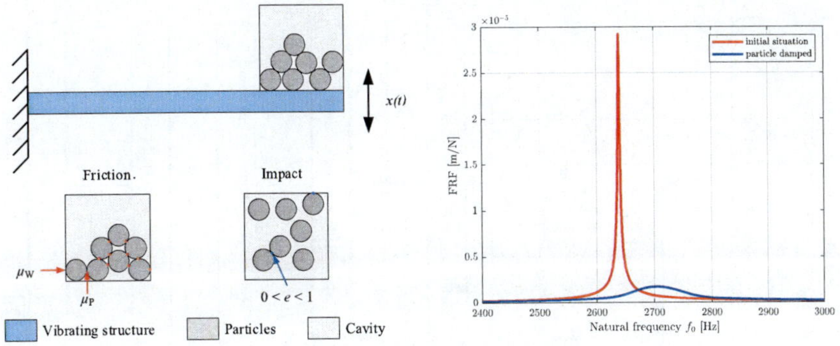

Fig. 2. Schematic diagram of a particle damper [21]

3 Design of a motorcycle triple clamp

In order to reduce critical hand-arm vibrations on the motorcycle, a motorcycle triple clamp is designed in this chapter under the aspects of high stiffness and high damping with low mass at the same time. In the first step a method for the design of stiffness and damping optimised components is presented. Afterwards the effect of particle damping is quantified by a parameter study. Based on this, the loads at the triple clamp are identified and the developed method is applied to the motorcycle triple clamp. For this purpose, the stiffness to weight ratio of the triple clamp is optimised by means of a topology optimisation. Subsequently, cavities for particle damping are integrated in the vibration-prone areas of the triple clamp, which are identified from a modal analysis. Finally, a static simulation is carried out to verify the mechanical strength.

3.1 Methological approach

The procedure for designing a stiffness and damping optimised component is shown in the form of a flow chart in Fig. 3 and is described in detail below. In order to solve conflicts of objectives during component design with regard to high stiffness combined with high damping, particle damping must be selected as a suitable measure. For example, the particles can be integrated in the area of the neutral fibre, thus enabling a component design optimised in terms of stiffness and damping.

Subsequently, the critical mode shapes, natural frequencies and occurring forces must be identified and calculated, since the effect of particle damping is highly non-linear [21,44,45] and is influenced by it. As there are usually no damping values for these boundary conditions, they must be quantified. For this purpose simulation methods like the Discrete Element Method (DEM) or experiments like a parameter study can be performed. For a parameter study the set-up has to be chosen in a way that the load cases (tension, compression, bending or superposition) correspond to those of the structural component to be optimised. By means of the parameter study, damping curves are determined as a function of different excitation forces, natural frequencies and cavity volumes, so that the damping of the structural component to be optimised can be estimated.

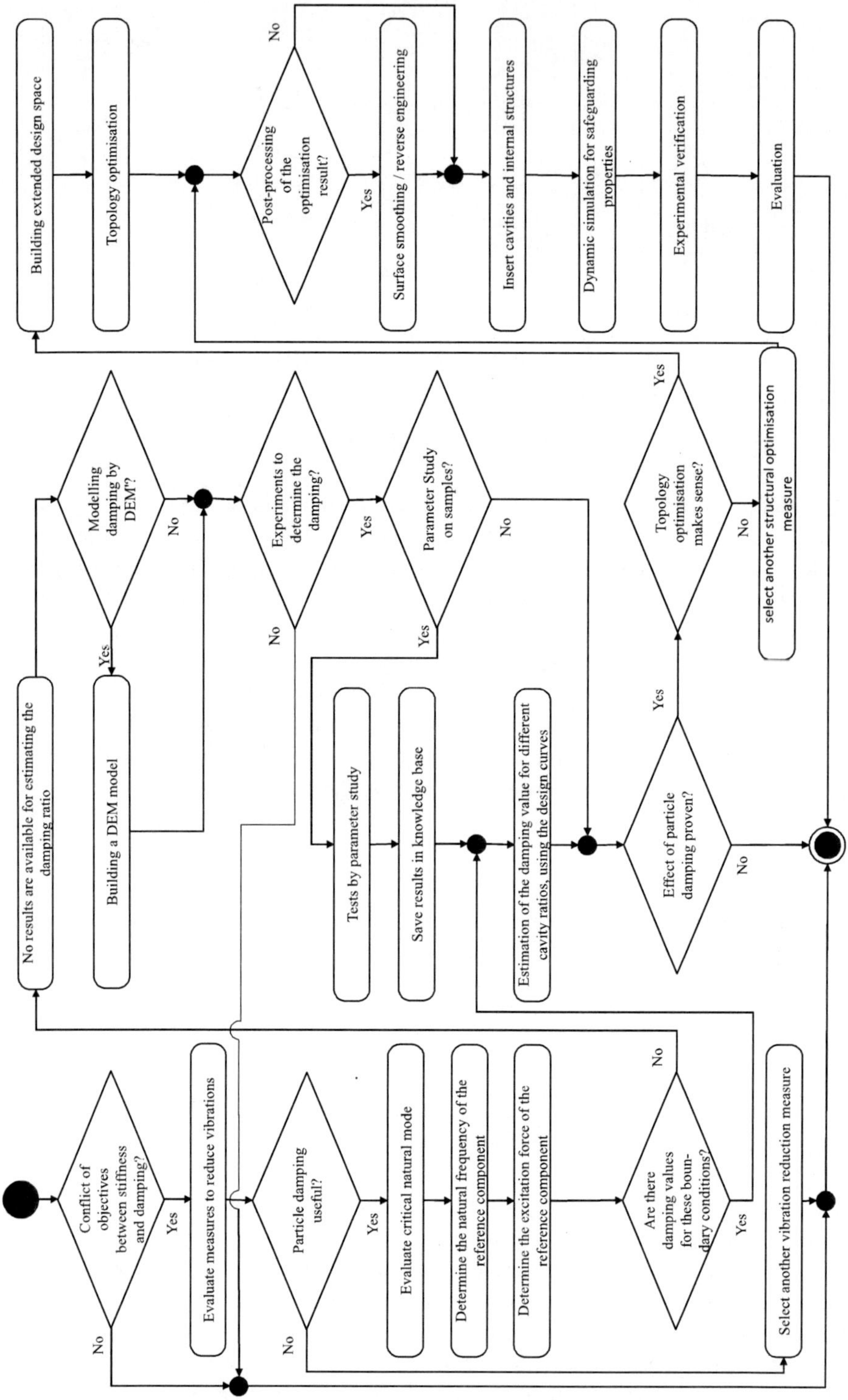

Fig. 3. Flow Chart for designing stiffness and damping optimised components, based on [21]

After the effect of particle damping has been proven and quantified, the desired structural component can be designed with regard to stiffness and damping. In the first step the stiffness to weight ratio is improved by structural optimisation, e.g. topology optimisation. In particular for the results of a topology optimisation, which are usually available in STL format, a surface smoothing and reverse engineering must be carried out. Based on this, cavities can be integrated into the structural component. For larger cavities, internal structures can sometimes be used to support overhangs or provide additional stiffness. The cavities should be integrated in the areas of the highest amplitudes that can be determined from a modal analysis. The desired degree of damping can be determined by the size of the cavity, which can be estimated from the previously determined damping curves (either from the DEM or parameter study). Subsequently, the component lifetime has to be verified simulatively. Finally, the structural component must be manufactured and the component properties must be compared to the requirements by means of test bench trials.

3.2 Determining the load conditions on the motorcycle stem

In this work a motorcycle triple clamp of a Ducati Monster 1100S is optimised. The motorcycle triple clamp is part of the stem and has a decisive influence on the driving stability. The corresponding CAD data of the original model are available under [46]. Altogether, the three load cases constant speed, cornering and emergency braking are considered for the design of the triple clamp. In order to determine the contact forces at the triple clamp, an FE model of the front wheel guide is constructed.

For the calculation of the braking force, the maximum case is assumed that the rear wheel in particular loses contact with the road surface and the entire weight force acts on the front wheel. In addition, it must be ensured that the motorcycle does not roll over due to braking too hard. Thus, the braking force can be calculated according to eq. 1 [47,48]. It should be noted that a possible air resistance F_w during braking was neglected.

$$F_{b\_max} = \min[\mu \cdot m \cdot g \, ; \, m \cdot g \cdot (p-b)/h] \tag{1}$$

The friction coefficient between tyre and road can be assumed to be $\mu = 1.4$ [23,47,48]. This can be achieved under optimal conditions, on dry and non-slip asphalt as a road surface and with an appropriate tyre mixture [23]. The maximum permissible total weight of 390 kg is defined as the total vehicle mass m. p represents the wheelbase and b and h the distance between the rear wheel and the centre of gravity in the horizontal and vertical directions respectively (see Fig. 4).

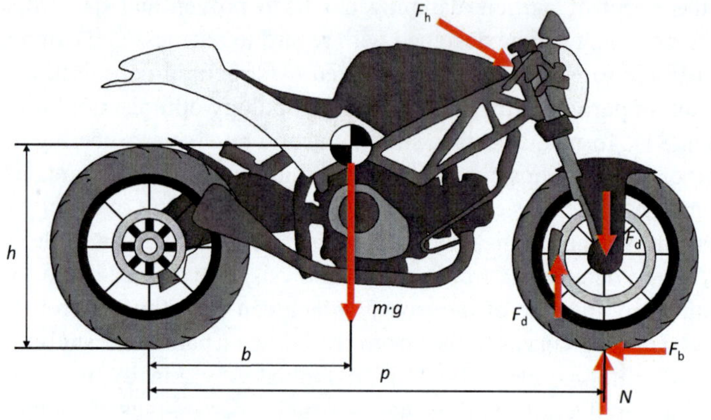

Fig. 4. Forces on the motorcycle during emergency braking

This results in the following expression for the braking force:

$$F_{b,max} = \min[1{,}4{\cdot}390\ kg{\cdot}9{,}81\ m/s^2\ ;\ 390{\cdot}9{,}81{\cdot}(1450\ mm{-}625\ mm)\ /\ 650\ mm] \quad (2)$$

$$F_{b,max} = \min[5356\ N;\ 4855\ N] = 4855\ N \quad (3)$$

When the brakes are applied to the brake disc, a force F_d, is generated, which can be calculated as follows [47–49]:

$$F_d = F_{b,max}{\cdot}d_w\ /\ d_d = 4855\ N{\cdot}570\ mm\ /\ 286\ mm = 9678\ N \quad (4)$$

Here d_w represents the diameter of the wheel and d_d the diameter of the brake disc. Due to equilibrium conditions, the force F_d acts both on the brake disc and on the wheel axle.

In addition to the braking force on the wheel, the rider's arm force with which he or she supports himself or herself on the handlebars must be considered during emergency braking. The arm position is of decisive importance for the arm power. For a relaxed sitting position on the bike, an angle α = -30° is assumed before the emergency stop, which indicates the position of the handlebar grip in relation to the horizontal plane through the shoulder joint. In addition, the angle β = 30° is assumed, which indicates the position of the handlebar grip in relation to the median plane of the driver's body. The distance of the handlebar grip l is set at 75 % of the maximum arm reach. According to Rohmert a standing man is able to apply a force of 186 N per arm under these boundary conditions [50]. Since in normal situations a man can only apply 30 – 40 % of the absolute force, the absolute force must be taken into account during emergency braking [51]. A safety factor of s = 2 is also applied.

$$F_{handlebar} = F_{hand}\ (\alpha, \beta, l)\ /\ \eta{\cdot}s = 186\ N\ /\ 0{,}3{\cdot}2 \approx 1250\ N \quad (5)$$

When cornering, the total weight of the motorbike is loaded on both tyres and must be divided and can be calculated according to eq. 6.

$$F_{K,y} = m{\cdot}g\ \mu\ b/p = 390\ kg{\cdot}9{,}81\ m/s^2{\cdot}1{,}4\ 625\ mm/1450\ mm = 2308\ N \quad (6)$$

For the force $F_{K,z}$ in vertical direction the unevenness of the road surface must be taken into account. On European roads, the overload factor x is usually between 3 and 5 [23]. In the following, an overload factor of 3 is assumed, because if the overload factor is too high, the road contact would be lost during cornering.

$$F_{K,z} = m \cdot g\ b/p\ x = 390\ \text{kg} \cdot 9{,}81\ \text{m/s}^2 \cdot 625\ \text{mm}/1450\ \text{mm} \cdot 3 = 4947\ \text{N} \quad (7)$$

For constant driving straight ahead, an overload factor of $x = 5$ is taken into account due to unevenness of the road surface. Air and rolling resistance play a minor role and can be neglected.

$$F_z = m \cdot g\ b/p \cdot x = 390\ \text{kg} \cdot 9{,}81 \cdot 625\ \text{mm}\ /\ 1450\ \text{mm} \cdot 5 = 8245\ \text{N} \quad (8)$$

3.3 Parameter study of the triple clamp

The dynamic properties of particle damping are determined by a Design of Experiment (DoE) by investigating the relevant parameters natural frequency, excitation force and cavity volume. The DoE is performed in the form of a parameter study (26 parameter combinations) using laser beam melted bending beams, since it is assumed that the critical vibration modes of the motorcycle triple clamp resemble a bending mode. In addition, AlSi10Mg was chosen as the material, as the original model of the triple clamp was cast from high-strength aluminium.

To carry out the test, the beams are supported on foam and excited to oscillate by means of an automated impulse hammer. The vibrations are measured by an acceleration sensor and the damping is calculated using the Circle-Fit method. Fig. 5 shows an example of a frequency transfer function (FRF) for three 5x5x200 mm beams. Here a broadband amplitude reduction of 600 - 18000 Hz can be seen for the particle-damped beams. The particle damped beams have a cavity of 10 % or 20 % of the total beam volume. Thus, even in filigree structures the effect of particle damping can be used if the cross-section of the cavity is larger than $A = 3\ \text{mm}^2$. Further test series show that the damping can be increased by up to a factor of 20 and are described in detail in [21]. In total, the beams are excited to vibrations from 10 to 180 N, thus characterising the frequency- and force-dependent damping curve.

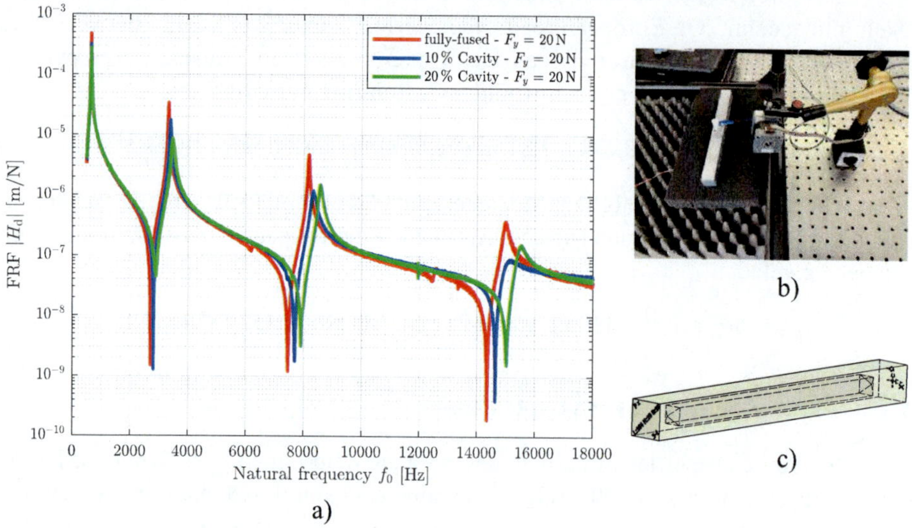

Fig. 5. a) Frequency Response Function of three 5 x 5 x 200 mm³ beams, b) Automated impulse hammer excitation of the beams, c) CAD model of the particle-filled beams [21]

3.4 Topology optimisation

To optimise the stiffness and mass of the motorcycle triple clamp, a topology optimisation is carried out. For this purpose, the extended design space is built up as a first step and is shown in Fig. 6. Thereby, attention was paid to the high level of functional integration. For the topology optimisation the force reactions are determined in *Ansys Workbench 2020* from an FE model of the motorcycle stem at the mounting points of the telescopic forks and the handlebars. For this purpose the load conditions of constant driving, cornering and emergency braking are taken into account. Due to the high forces of sometimes more than 5 kN in different load conditions, high component stresses result which exceed the permissible stresses of AlSi10Mg, so that the steel 1.2709 is selected as material for the motorcycle triple clamp and used for the topology optimisation. The mass target for the optimisation is 1 kg. The result of the topology optimisation is shown in Fig. 6. Here it can be seen that the surfaces still have to be smoothed and prepared. Using the *Retopology-Tool* from *Autodesk AliasAutoStudio2021*, the geometry could be reconstructed within a few hours. The functional surfaces are then detailed in *Autodesk Inventor*. The reconstructed result from the topology optimisation is shown in Fig. 6.

In total 21 individual components of the original triple clamp are brought together. These include the triple clamp, handlebar mount, damping elements, cockpit brackets and various mounting elements.

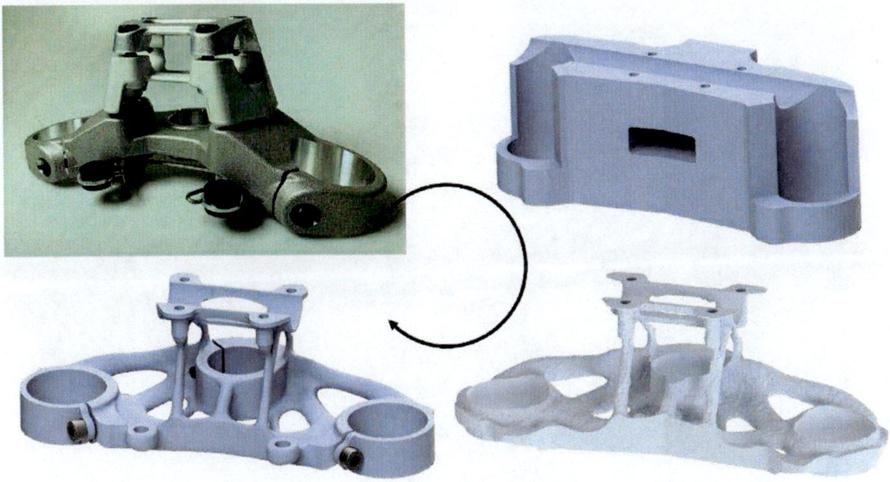

Fig. 6. Steps of topology optimisation for the motorcycle triple clamp

3.5 Insertion of cavities for the integration of particle damping

To integrate the effect of particle damping into the triple clamp, a modal analysis is carried out in the frequency range up to 2500 Hz and the areas with high vibration amplitudes are evaluated. The modal analysis is carried out for the motorcycle stem so that the influence of neighbouring components on the vibration modes and natural frequencies is taken into account. Two vibration modes are shown in Fig. 7. It is noticeable that the vibrations occur primarily at the mountings of the telescopic forks and handlebars. Accordingly, the cavities are placed in these areas (see Fig. 8). In addition, particle-filled cavities are provided in the areas of the cockpit mount and the steering head bearing.

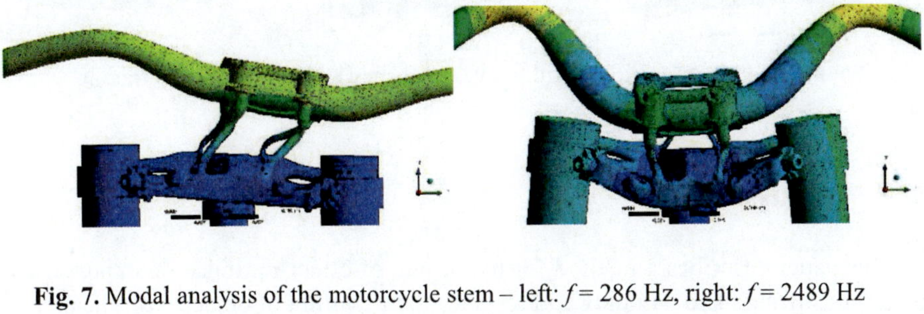

Fig. 7. Modal analysis of the motorcycle stem – left: $f = 286$ Hz, right: $f = 2489$ Hz

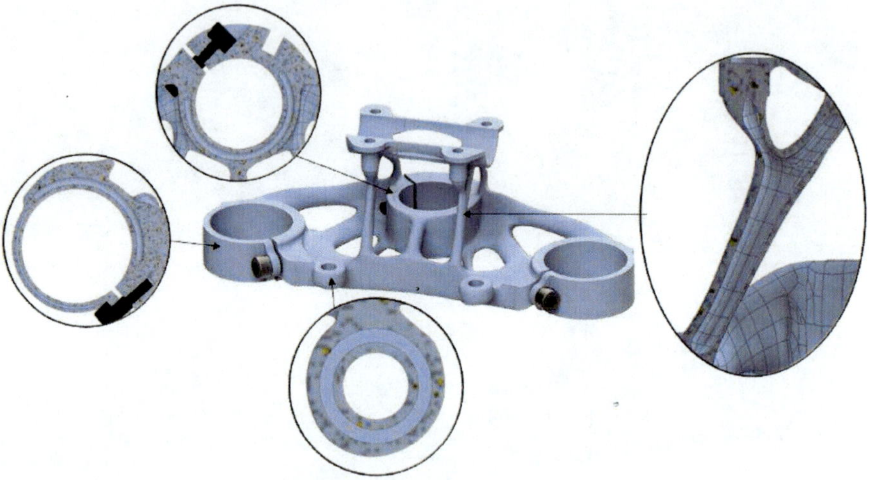

Fig. 8. Integration of particle-filled cavities in the motorcycle triple clamp

3.6 Proof of static strength

A static simulation is carried out to prove the strength of the triple clamp with cavities under the maximum loads. The stress simulation has shown that the alloy 1.2709 has sufficient strength, as this material can withstand 10^6 load cycles up to 400 N/mm² and is therefore fatigue resistant [52,53]. An aluminium alloy could not have withstood the corresponding loads. Fig. 9 shows the stress analysis for the three load conditions emergency braking, cornering and constant driving.

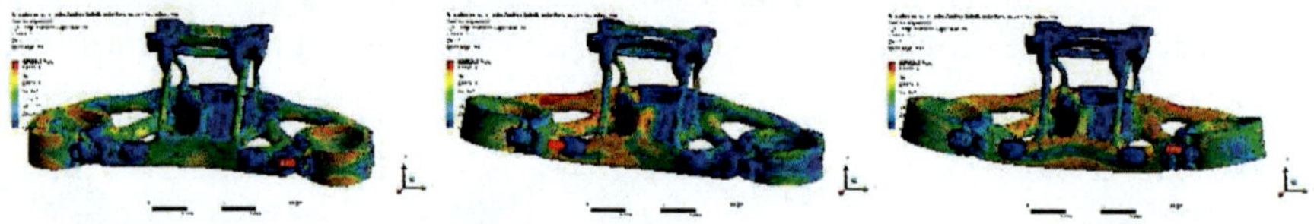

Fig. 9. Proof of static strength - left: emergency braking, middle: cornering, right: constant driving

4 Discussion

This paper describes a method for the design of effect-optimised components with regard to stiffness and damping and resolves their conflict of objectives. The presented method was demonstrated by means of the application example of a motorcycle triple clamp. The first part of the method, the stiffness and strength design, could be successfully proven by simulation. The transferability of the damping values from a parameter study to the demonstrator has not been carried out at this time, but will be investigated in further work. Furthermore, the effect of particle damping for the load conditions torsion, tension/compression and their combination can be quantified in further work, which was not the subject of this paper.

In addition, it is conceivable to adapt this method for a multi-criteria effect optimisation. For example, for flow-optimised thin-walled geometries with external force application, such as aerospace components. The multi-criteria effect optimisation may become more and more important in the future, since the continuous development of the LPBF system technology allows the production of larger and larger components. In the case of large components, it will be of great interest to integrate the damping elements directly into the interior of the component in order to push the design limits and increase the degree of functional integration and component performance.

A further potential could be raised if the presented method is supported by a computer aided engineering environment. Up to now, a variety of different software is required, starting with the characterisation of the damping from the parameter studies, FEM simulation up to the reconstruction. If the interface problems are overcome, for example, the free-form geometries of the cavities in the structural component could be built up in a parameterised way. In this way, the size of the cavity could be varied by means of a simulative parameter study and the structural component could be evaluated in terms of stiffness, damping and strength within a very short time. At the same time, the manufacturing restrictions of additive manufacturing could be permanently checked. It would also be possible to use a generative design approach for the creation of the cavities. Through this interaction and interlocking of the computer tools, the component could be better adapted to the requirements.

From the results of the motorcycle triple clamp it can be concluded that the effect of particle damping is particularly suitable for components that are designed with regard to the conflicting objectives of stiffness vs. damping. From static simulation results for the motorcycle triple clamp, it could be determined that the strength remains approximately constant after the insertion of cavities. This is particularly interesting for components that are subjected to bending stress, so that the cavities are integrated in the areas of the neutral fibre (Damping for free). However, dynamic test bench trials still have to be carried out here for evaluation.

An estimation of the damping values for the motorcycle triple clamp is difficult because the steel alloy 1.2709 was chosen as material for the triple clamp due to the high stresses. Therefore, the results of the parameter study for the AlSi10Mg alloy can only be transferred to a limited extent. For this purpose, a parameter study for 1.2709 will be carried out in further work.

Due to the change of material, the triple clamp has become approx. 20 % heavier compared to the original model, from $m_0 = 1.286$ kg to $m_{opt} = 1.55$ kg. However, sometimes when good damping in the triple clamp is demonstrated, damping elements at other points in the motorcycle can be simplified or omitted. This could save further weight, assembly work and costs.

Alternatively, the assumptions regarding the high loads on the triple clamp can be quantified by reverse engineering and thus sometimes be lower. This would make it possible to switch to an aluminium or titanium alloy.

5 Conclusion

In this paper a method for the design of effect optimised components was presented under the aspect of high stiffness combined with high damping. The core of this method is the use of particle damping as a measure for vibration reduction. The effect of particle damping is quantified with the help of a parameter study within the framework of a DoE, as this is highly non-linear and time-consuming to simulate, thus ensuring transferability to real structural components. The effect of particle damping is proven for the AlSi10Mg alloy from 600 to 18000 Hz and in some cases the damping can be increased by a factor of more than 20.

The developed method for the design of effect optimised components regarding stiffness and damping is illustrated by the application example of a motorcycle triple clamp. By means of laser powder bed fusion, the particle-filled cavities can be integrated directly into the interior of the component in the low-stress areas of the component, around the neutral fibre. The insertion of particle-filled cavities could be realised even with topology optimised results. The static strength could be proved simulatively for the material 1.2709.

Due to the high non-linearity, it is difficult to explicitly evaluate the effect of particle damping for the triple clamp, as it depends on force, natural frequency, natural shape and cavity volume. On average, however, a damping increase of factor 3-5 is expected for different bending modes if the triple clamp would be made of AlSi10Mg.

6 References

1. Wiberg, A., Persson, J., Ölvander, J.: Design for additive manufacturing – a review of available design methods and software. Rapid Prototyping Journal 25(6), 1080–1094.
2. Lachmayer, R., Lippert, R., B.: Entwicklungsmethodik für die Additive Fertigung. Springer Berlin Heidelberg, Berlin Heidelberg (2020).
3. Kumke, M.: Methodisches Konstruieren von additiv gefertigten Bauteilen. Springer Fachmedien Wiesbaden, Wiesbaden (2018).
4. Diegel, O., Nordin, A., Motte, D.: A Practical Guide to Design for Additive Manufacturing. Springer Singapore, Singapore (2019).
5. Lachmayer, R., Bode, B., Grabe, T., Rettschlag, K.: Integration spezifischer Effekte in Strukturbauteilen mittels additiver Fertigungsverfahren. In: Lachmayer, R., Rettschlag, K., Kaierle, S. (eds.) Konstruktion für die Additive Fertigung 2019, pp. 1–10. Springer Berlin Heidelberg, Berlin Heidelberg (2020).
6. Lachmayer, R., Gembarski, P., C., Gottwald, P., Lippert, R., B.: The Potential of Product Customization Using Technologies of Additive Manufacturing. In: Bellemare, J., Carrier, S., Nielsen, K., Piller, F., T. (eds.) Managing Complexity 2015, pp. 71–81. Springer International Publishing, Cham (2016).
7. Yang, S., Tang, Y., Zhao, Y., F.: Assembly-Level Design for Additive Manufacturing: Issues and Benchmark. In: ASME 2016 International Design Engineering Technical Conferences and Computers and Information in Engineering Conference, vol. 2A, American Society of Mechanical Engineers, Charlotte, North Carolina (2016).
8. Tang, Y., Yang, S., Zhao, Y., F.: Sustainable Design for Additive Manufacturing Through Functionality Integration and Part Consolidation. In: Muthu, S., S., Savalani, M., M., (eds.)

Handbook of Sustainability in Additive Manufacturing, pp. 101–144. Springer Singapore; Singapore (2016).

9. Ehlers, T., Lippert, R., B., Lachmayer, R.: Bewertung von Strukturbauteilen aus gradierten Materialien für Selektives Laserstrahlschmelzen. In: Lachmayer, R., Lippert, R., B., Kaierle, S. (eds.) Konstruktion für die Additive Fertigung 2018, pp. 109–127. Springer Berlin Heidelberg, Berlin Heidelberg (2019).
10. Scott-Emuakpor, O., George ,T., Runyon, B., Holycross, C., Langley, B., Sheridan, L., O'Hara, R. Johnson, P., Beck, J.: Investigating Damping Performance of Laser Powder Bed Fused Components With Unique Internal Structures. In: ASME Turbo Expo 2018: Turbomachinery Technical Conference and Exposition, pp. V07CT35A020. ASME, Oslo Norway (2018).
11. Scott-Emuakpor, O., George, T., Runyon, B., Langley, B., Sheridan, L., Holycross, C., O'Hara, R., Johnson, P.: Forced-Response Verification of the Inherent Damping in Additive Manufactured Specimens. In: Kramer, S., Jordan, J.,L., Jin, H., Carroll, J., Beese, A.,M. (eds.) Mechanics of Additive and Advanced Manufacturing, vol. 8, pp. 81–86. Springer International Publishing, Cham (2019).
12. Scott-Emuakpor, O., George, T., Runyon, B., Beck, J., Sheridan, L., Holycross, C., O'Hara, R.: Sustainability Study of Inherent Damping in Additively Manufactured Nickel Alloy. AIAA Journal 57, 456–461 (2019).
13. Scott-Emuakpor, O., George, T., Runyon, B., Holycross, C., Sheridan, L., O'Hara, R.: Assessing Additive Manufacturing Repeatability of Inherently Damped Nickel Alloy Components. J. Eng. Gas Turbines Power (2019).
14. Scott-Emuakpor O, Beck J, Runyon B, George T. Validating a Multifactor Model for Damping Performance of Additively Manufactured Components. AIAA Journal, 1–8 (2020).
15. Scott-Emuakpor, O., Schoening, A., Goldin, A., Beck, J., Runyon, B., George, T.: Internal Geometry Effects on Inherent Damping Performance of Additively Manufactured Components. AIAA Journal, 1–7 (2020).
16. Künneke, T., Zimmer, D.: Funktionsintegration additiv gefertigter Dämpfungsstrukturen bei Biegeschwingungen. In: Richard, H., A., Schramm, B., Zipsner, T. (eds.) Additive Fertigung von Bauteilen und Strukturen, pp. 61–74. Springer Fachmedien Wiesbaden, Wiesbaden (2017).
17. Schmitz, T., Betters, E., West, J.: Increased damping through captured powder in additive manufacturing. Manufacturing Letters 25, 1–5 (2020).
18. Schmitz, T., Gomez, M., Ray, B., Heikkenen, E., Sisco, K., Haines, M., Osborne, J., S.: Damping and mode shape modification for additively manufactured walls with captured powder. Precision Engineering 66, 110–124 (2020).
19. Vogel, F., A., Berger, S., Özkaya, E., Biermann, D.: Vibration Suppression in Turning TiAl6V4 Using Additively Manufactured Tool Holders with Specially Structured, Particle Filled Hollow Elements. Procedia Manufacturing 40, 32–7 (2019).
20. Ehlers, T., Lachmayer, R.: Einsatz additiv gefertigter Partikeldämpfer – eine Übersicht. In: Lachmayer, R., Rettschlag, K., Kaierle, S. (eds.) Konstruktion für die Additive Fertigung 2019, pp. 123–142. Springer Berlin Heidelberg, Berlin Heidelberg (2020).
21. Ehlers, T., Tatzko, S., Lachmayer, R., Wallaschek, J.: Design of particle dampers for additive manufacturing. Additive Manufacturing 38C, 101752 (2021).
22. Hopkinson, N., Dickens, P. M., Hague, R., J., M.: Rapid manufacturing: An industrial revolution for the digital age. John Wiley, Chichester England (2006).
23. Stoffregen, J.: Motorradtechnik. 9th edn. Springer Fachmedien Wiesbaden, Wiesbaden (2018).

24. Tarabini, M., Mauri, N., Gaudio, I., Cinquemani, S., Moorhead, A., P., Bongiovanni, R., Feletti, F.: Hand-arm vibration in motocross: measurement and mitigation actions. Muscle Ligaments and Tendons J 10(2), 280-289 (2020).
25. Poole. C., J., M., Mason, H., Harding, A.-H.: The relationship between clinical and standardized tests for hand-arm vibration syndrome. Occupational Mededicine 66(4), 285-291 (2016).
26. ISO 5349-1: Mechanical vibration – Measurement and evaluation of human exposure to hand-transmitted vibration – Part 1: General requirements (2001).
27. Fasana, A., Giorcelli, E., A.: Vibration absorber for motorcycle handles. Meccanica 45(1), 79–88 (2010).
28. Bovenzi, M.: Exposure-response relationship in the hand-arm vibration syndrome: an overview of current epidemiology research. International archives of occupational and environmental health 71(8), 509-519 (1998).
29. Mirbod, S., M., Yoshida, H., Jamali, M., Masamura, K., Inaba, R., Iwata, H.: Assessment of hand-arm vibration exposure among traffic police motorcyclists. International archives of occupational and environmental health 70(1), 22-28 (1997).
30. Jelačić, Z., Pikula, B.: Vibration Analysis of Motorcycle Handles. In: Karabegović, I., (eds.) New Technologies, Development and Application, vol. 42, pp. 196–201. Springer International Publishing, Cham (2019).
31. Chen, H.-C., Chen, W.-C., Liu, Y.-P., Chen, C.-Y., Pan, Y.-T.: Whole-body vibration exposure experienced by motorcycle riders – An evaluation according to ISO 2631-1 and ISO 2631-5 standards. International Journal of Industrial Ergonomics 39(5), 708-718 (2009).
32. Noh, J., M., Rezali, K., A., M., As'arry, A., Jalil, N., A., A.: Transmission of vibration from motorcycle handlebar to the hand. Journal of Society of Automotive Engineers Malaysia. 1(3), 191–197 (2017)
33. Burg, H., Moser, A.: Handbuch Verkehrsunfallrekonstruktion. 3rd edn. Springer Fachmedien Wiesbaden, Wiesbaden (2017).
34. Baad, S., M., Patil, R., J., Qaimi, M., G.: Hand Arm Vibration Alleviation of Motor-cycle Handlebar using Particle Damper. International Journal of Engineering and Manufacturing 7(1), 26–40 (2017).
35. Ehlers, T., Lachmayer, R., Vajna, S., Halle, T.: Producibility. In: Vajna, S., (eds.) Integrated Design Engineering, pp. 287–323. Springer International Publishing, Cham (2020).
36. Friend, R., D., Kinra, V., K.: Particle Impact Damping. Journal of Sound and Vibration 233(1), 93-118 (2000).
37. Lu, Z., Wang, Z., Masri, S., F., Lu, X.: Particle impact dampers: Past, present, and future. Structural Control and Health Monitoring 25(1) e2058 (2017).
38. Papalou, A., Masri, S., F.: Performance of Particle Dampers Under Random Excitation. Journal of Sound and Vibration 118(4), 614–621 (1996).
39. Panossian, H., V.: Structural Damping Enhancement Via Non-Obstructive Particle Damping Technique. Journal of Vibration and Acoustics 114(1), 101–105 (1992).
40. Fowler, B., L., Flint, E., M., Olson, S., E.: Effectiveness and predictability of particle damping. In: Hyde, T., T. (eds.) SPIE's 7th Annual International Symposium on Smart Structures and Materials, pp. 356–367. CA: SPIE, Newport Beach (2000).
41. Olson, S.,E.: An analytical particle damping model. Journal of Sound and Vibration 264(5), 1155-1166 (2003).
42. Saeki, M.: Analytical study of multi-particle damping. Journal of Sound and Vibration 281(3-5), 1133–1144 (2005).

43. Xiao, W., Li, J., Wang, S., Fang, X.: Study on vibration suppression based on particle damping in centrifugal field of gear transmission. Journal of Sound and Vibration 366, 62-80 (2016).
44. Mao, K., Wang, M.,Y., Xu, Z., Chen, T.: Simulation and Characterization of Particle Damping in Transient Vibrations. Journal of Sound and Vibration 126(2), 202 -211 (2004).
45. Fowler, B.,L., Flint, E., M., Olson, S., E.: Design methodology for particle damping. In: Inman, D.,J. (eds.) SPIE's 8th Annual International Symposium on Smart Structures and Materials, pp. 186–197. SPIE, Newport Beach (2001).
46. Kronslev, K.: Ducati Monster 1100S, https://grabcad.com/library/ducati-monster-1100s. last accessed 2020/11/23.
47. Croccolo, D., De Agostinis, M., Vincenzi, N.: Analytical, Numerical and Experimental Study of the Effects of Braking on Single Disc Motorcycle Forks. In: Öchsner, A., Da Silva, L., F., M., Altenbach, H. (eds.) Materials with Complex Behaviour II, vol. 16, pp. 51–79, Springer Berlin Heidelberg, Berlin Heidelberg (2012).
48. Croccolo, D., De Agostinis, M.: Motorbike Suspensions. Springer London, London (2013).
49. Croccolo, D., De Agostinis, M., Vincenzi, N.: An analytical approach to the structural design and optimization of motorbike forks. Proceedings of the Institution of Mechanical Engineers, Part D: Journal of Automobile Engineering 226(2), 158–168 (2012).
50. Rohmert, W.: Maximalkräfte von Männern im Bewegungsraum der Arme und Beine. pp. 30. VS Verlag für Sozialwissenschaften, Wiesbaden (1966).
51. Fritzsche, J.: Selbstverteidigung. In: Heimann, R., Fritzsche, J. (eds.) Gewaltprävention in Erziehung, Schule und Verein, pp. 205–228. Springer Fachmedien Wiesbaden, Wiesbaden (2020).
52. Tshabalala, L., Sono, O., Makoana, W., Masindi, J., Maluleke, O., Johnston, C., Masete, S.: Axial fatigue behaviour of additively manufactured tool steels. Materials Today: Proceedings 2020 (2020).
53. Croccolo, D, De Agostinis, M., Fini, S., Olmi, G., Robusto, F., Ćirić Kostić, S., Vranić, A., Bogojević, N.: Fatigue Response of As-Built DMLS Maraging Steel and Effects of Aging, Machining, and Peening Treatments. Metals 8(7), 505 (2018).

Fully Automated Subdivision Surface Parametrization for Topology Optimized Structures and Frame Structures using Euclidean Distance Transformation and Homotopic Thinning

Martin Denk[1[0000-0002-0204-3608]] Klemens Rother[2[0000-0002-9643-4967]] Kristin Paetzold[1[0000-0002-8638-3713]]

[1] Bundeswehr University Munich
Institute for Product Development, 85577 Neubiberg, Germany
[2] Munich University of Applied Science
Institute of Material and Building Research, 80335 Munich

Keywords: Topology Optimization, Reverse Engineering, Parametrization, Subdivision Surfaces

Abstract

Polygon meshes and particularly triangulated meshes can be used to describe the shape of different types of geometry such as bicycles, bridges, or runways. In engineering, such polygon meshes can be supplied as finite element meshes, resulting from topology optimization or from laser scanning. Especially from topology optimization with low member size settings, frame-like polygon meshes with slender parts are typical and often have to be converted into a CAD (Computer-Aided Design) format, e.g., for further geometrical detailing or performing additional shape optimization. Especially for such frame-like geometries, CAD designs are constructed as beams with cross-sections and beam-lines, whereby the cross-section is extruded along the beam-lines or beam skeleton. In our research, automatic parameterization of polygon meshes into a subdivision surface representation is tried out. For this purpose, the beam-lines are approximated by computation of curved skeletons, which are determined by a homotopic thinning method. These skeleton lines are transformed into a subdivision surface control grid by using the Euclidean distance transformation.

1 Introduction

Various applications such as animation design, topology optimization, or 3D laser scanning apply discrete triangulated meshes to represent the surface of a 3D geometry. Triangulated surfaces describe the shape of the object by separating the boundary surface into discrete triangles. Engineering applications for manufacturing, optimization, or design modification require a parametric representation such as constructive solid geometry (CSG) or free form surfaces of the geometry [1], [2]. The boundary representation of an object can be reconstructed with primitive surface types [1]–[3]

S. Pfingstl et al. (eds.), *Proceedings of the Munich Symposium on Lightweight Design 2020*, https://doi.org/10.1007/978-3-662-63143-0_2

or free form surfaces [4]–[6], which is called boundary representation surface (Brep). Primitive surface types are typically used for (CSG) [1]–[3], whereas free form surfaces are used for the reconstruction of organic geometries [4]–[6].

The manual parametrization of these polygon meshes can be time-consuming and can lead to inaccurate approximations or altering the topology. Geometric reverse engineering attempts to redesign a parametric shape and topology of representations such as polygon meshes [1]–[3] or volumetric geometries [7]. Most of the approaches perform a so-known mesh segmentation [8], where the boundary of the geometry is split into several patches [1]–[3]. Afterward, the surface of the geometry can be parametrized in surface types like primitive surfaces [1]–[3] or nonuniform rational B-spline (NURBS) surfaces [4], [5]. This segmentation only results in surface parametrization [1]–[3] so that the topology itself is not covered quite well. Therefore so-called curve (middle line) or surface skeleton (middle surface) can be used for part segmentation [8]–[10], which can lead to a more beam-line representation [7], [11], [12]. These curve skeletons serve as a reasonable shape descriptor for tube-like organic geometries [13]. Such geometries can be derived in topology optimization for linear elasto static, heat transfer, or considering both [14].

In our approach, a homotopic thinning method from the authors of [15] is chosen to determine the skeleton of 3D volumetric images, which ensures topology preservation. The authors of [11], [12], [16] uses parametric cross-sections, whereas we rather use a parameterization directly at each skeleton point that defines a radius, which is derived from the underlying concept of medial balls [17]. Medial balls are an alternative shape description that uses maximum in-described balls which are defined by the center of the ball and the radius [13], [18]. Therefore, for curve skeleton, a tube-shape polygon mesh should be applied [13].

Based on that, we perform a Euclidean distance transformation, which serves as an approximation for the boundary distance and thus for a correlating radius of each skeletal point. The following figure shows the desirable parameterization using a skeleton and a corresponding skin representation of the polygon mesh. By moving two skeleton nodes upwards (orange arrows), the underlying parameterization serves for shape adjustment. Additionally, the shape around each point can be scaled by a corresponding radius.

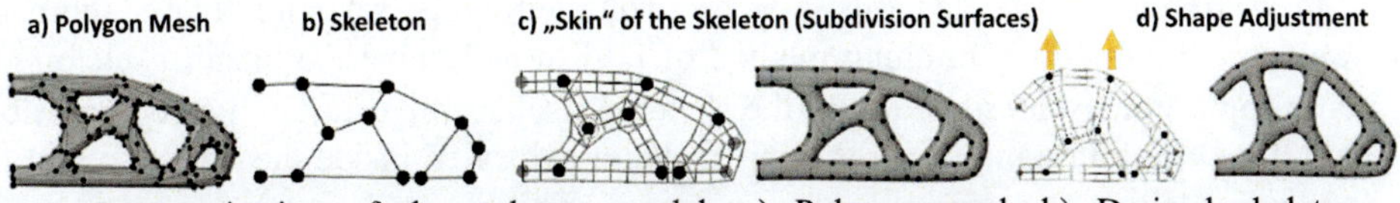

Fig. 1. Parametrization of the polygon model: a) Polygon mesh b) Derived skeleton c) Subdivision surface control grid resulting from the skeleton and the radius d) Shape adjustment by changing the skeleton

Subdivision surfaces divide the control grid by an iterative refinement [19]. This parameterization can be used for different tasks such as shape optimization [20], shape reconstruction, and reverse engineering [19], [21]. Our main contribution will be the fully automated of polygon models in topology optimization into a skin-based subdivision surface representation so that the subdivision control grid is bounded to the skeleton.

2 State of the Art

A 3D segmentation for CAD applications can be achieved by various methods such as spectral analysis, clustering, region growing, skeleton-based methods [8] or applying machine learning [22], [23]. The state of the art of this article refers to the skeleton-based methods, where first, the skeleton is extracted, and then the cross-sections are determined. Another option is to determine first the cross-section and then to construct the skeleton [24], which is not further discussed in this article. For other concepts, we refer to the summary of segmentation methods in [8] or more recent publications [2], [10], [22], [23], [25].

Typically, the skeleton is used to approximate the beam-line or the extrusion direction. Therefore, different skeletonization methods, such as contraction or homotopic thinning, can be used. In the recent publications of [11], [12], the beam-line is approximated by a contraction method, which is presented in [26]. Contraction methods act on the polygon mesh and iteratively push the boundary until the surfaces intersect each other [27]. These can result in a 1D curve or 1D/2D curve/surface skeletons. The contraction method is only applied to the boundary of the geometry, whereas thinning methods erode a volumetric representation. In a recent article, the authors in [28] uses directly the medial axis transformation for the estimation of curve/surface skeletons, which can despite to the contraction method of [26], fully cover the original shape [13]. These skeletons often cover the noise of the boundary, so that pre and post-processing steps is often required [13].

The authors of [7] use a thinning based approach for the calculation of the skeleton for 2D use cases. The topology optimization is performed on pixel elements and can therefore be processed with methods of digital image processing. In a more recent article of [29], the authors use the homotopic 3D thinning method of [15] for 3D cases. This method offers the advantage to ensure correct topological representation in the curve skeleton if no cavities are located in the object. The thinning method is applied to a solution of a topology optimization on finite elements represented as voxels. Afterward, these finite elements serve as a volumetric representation. In our approach, we use the homotopic thinning method of [15] to determine the binary skeleton and ensuring topology preservation in the skeleton itself, similar to the method described in [29]. During thinning, the resulting skeleton is located inside the object so that the endpoints have to be pushed back using cross-sections to cover the shape. So despite the authors of [29], we extend the endpoints on open branches until the boundary is hit. If a beam-line representation is established, the corresponding cross-section can be determined. The authors of [12] use manually defined elliptical and spherical cross-sections. Despite the manual selection, the authors of [11], [28]–[30] automatically uses shape properties such as radius of circular cross-sections [11], [30] or the local radius [28], which can be calculated using the Euclidian distance between the triangulated mesh and a corresponding skeleton segment [11], [30] or the radius of the medial ball [17], [28]. Alternatively, the radius can be approximated by an additional size optimization [29]. Using the beamline and the corresponding cross-sections the authors of [11], [12], [29], [30] extrude the cross-section along the skeleton-curves using cylindrical primitives [11], [29], [30] or elliptical cross-sections [12]. For a smooth surface, these extruded cross-sections can be furthermore blended [12] or combined with spheres on the cylindrical parts [29]. Despite the primitive cross-section of [11], [12], [29], the authors of [16] manually determine surface skinning with B-spline cross-sections to cover the shape of the topology optimization. Therefore, the whole smooth geometry is directly defined by the control grid and does not require any blending processing. In our approach, we use a skin representation similar to [16]. However, we rather use subdivision surfaces instead of B-Splines due to its flexible topology of nonequal number control grids along a surface segment [20], [31]. The cross section parametric surface is approximated similar to the methods in [11], [30], by applying the distance transformation on the binary image.

3 Parametrization

To fully parametrize an object, the presented method uses a homotopic thinning method, which requires a digital image. The following figure shows the steps from the rasterization to the parametric model acquisition. The steps are separated into action flow and data flow.

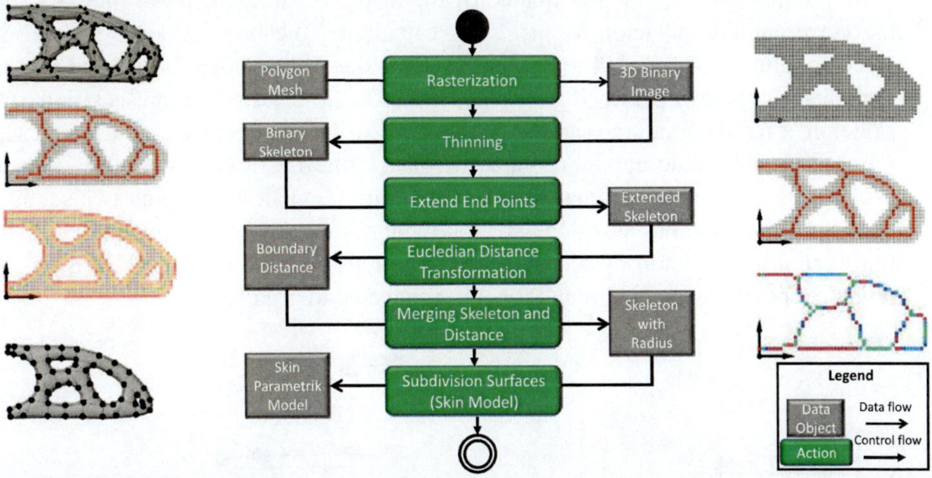

Fig. 2. Parametrization Steps

In the first step, the polygon mesh of the topology optimized model is rasterized into volumetric pixels (Voxels) of a 3D binary image using the implementation [32] of the method [33]. On that image, a homotopic thinning method of [15] is applied so that a skeleton representation (red voxels) is acquired. The endpoints on open branches are extended towards the boundary of the model. Based on the volumetric representation, a Euclidian distance transformation is calculated, which serves as a radius for the parametrization. Then, the corresponding radius is mapped to the points of the skeleton, where red voxels represent a small radius and blue voxels a large radius. These points and the corresponding radii are transferred to a skin subdivision surface representation. The following figure shows the iterative of the subdivision of a cube control grid where the rectangle grid is iteratively subdivided into a sphere like object.

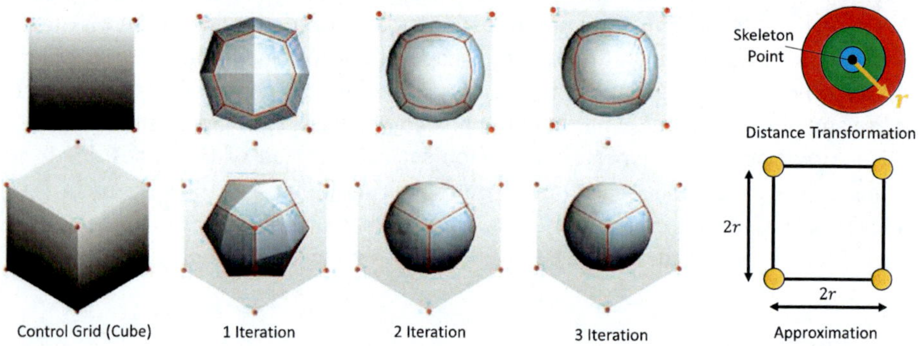

Fig. 3. Subdivision Surfaces of a cube and the approximation of the control grid by the radius

The control grid is approximated with the value from the distance transformation of the corresponding skeleton point. The rectangle length and width are two times the size of that distance value.

While the control grid is estimated by the distance value, the beam-lines are set to the corresponding skeleton segment. By calculating the line curvature change of a fitted B-spline based on [34], [35], each binary skeleton segment is reduced to a few numbers of topology preserving line segments. Due to the digital representation of the skeleton, a fixed curvature criterion can be chosen. In our use cases, we use as a separation criterion a line curvature value of one. If the maximum curvature along the skeleton segment is higher than these criteria, the segment is split into two segments on the highest curvature position. The following figure shows the estimation of beam-lines using the skeleton by analyzing the curvature and the topology. Each junction point, endpoint, and an additional point is connected to a skeleton.

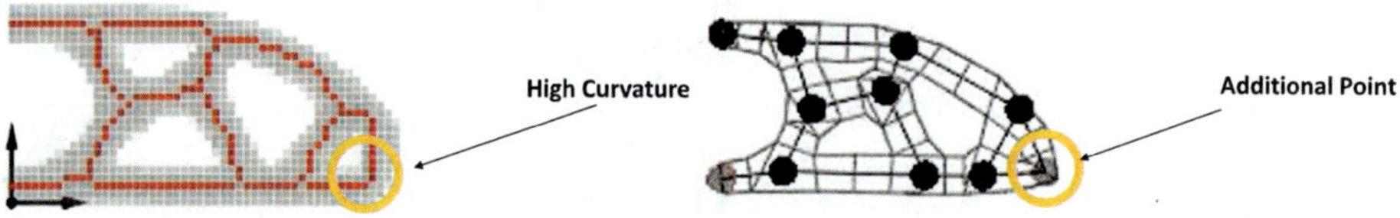

Fig. 4. Estimating beam-lines and inserting additional points on high line curvature

Straight skeleton segments are reduced to lines with one start and one endpoint. The marked segment in **Fig.** ***4*** is separated into two segments due to the high curvature. In the following chapter, the proposed method is tested on several different topology optimized objects and one train/bus frame.

4 Parametrization of Topology Optimization Results

The experimental part is separated into two sections, fully automated subdivision surface parametrization and shape adjustments on the skin subdivision surface. In the first section, the parametrization of the polygon mesh into subdivision surfaces is covered. The second section covers the flexible shape adjustment using subdivision skin shape modification. The visual examples show the rasterized model, the extracted chains, the distance transformation, and the corresponding skin parametrization. To cover the advantages and disadvantages of this method, we present examples that fulfill the necessary property for curved skeletons of the underlying concept of medial balls [13], [17], and some with a thin-walled shape.

4.1 Fully Automated Skin Subdivision Surface Parametrization

The following figure shows the five rasterized examples followed by a distance approximation. The thin-walled samples (2,3,4) shows a small thickness variation due to the thin representation so that the thickness values are similar (red). Sample one and five shows a thickness variation where blue represents a high and red a small thickness. Afterward, the rasterized object is skeletonized and split into several skeleton segments. These segments represent the beam-line with one start and one endpoint. Based on these segments and the distance, the skin parametric can be built by using the diameter as the length for the quadratic control grid.

Table 1. Subdivision surface parametrization in topology optimization

Use Case	Rasterized	Distance	Skeleton Segments	Skin Parametric	Subdivision Surface
1				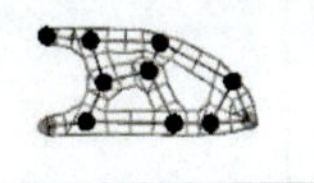	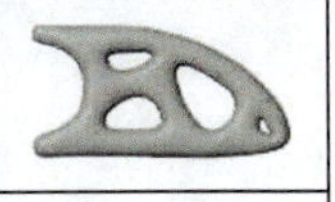
2	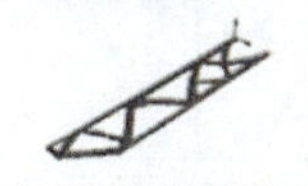	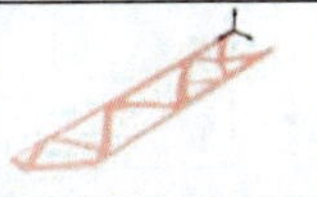	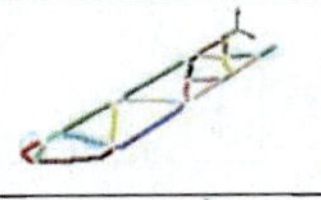	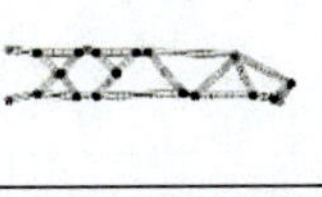	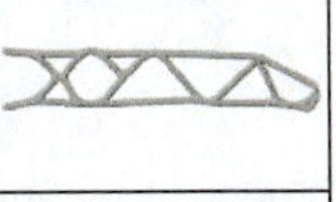
3			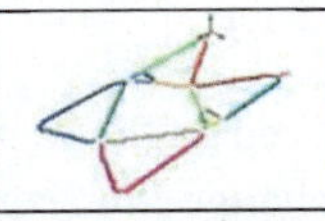	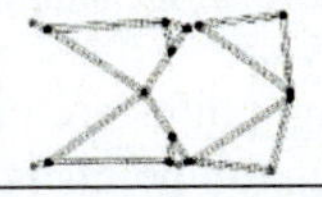	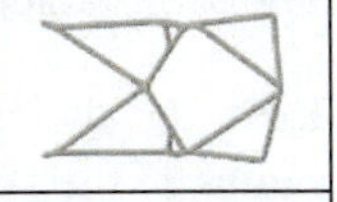
4			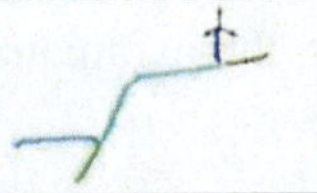	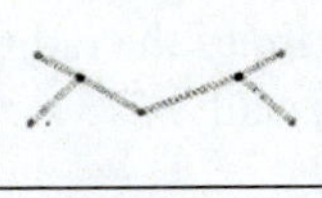	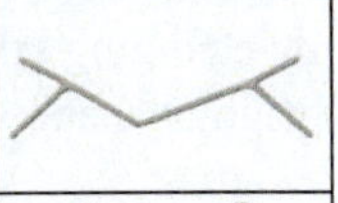
5				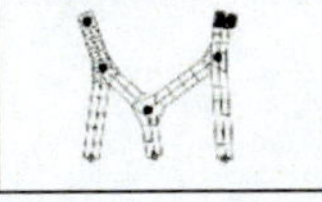	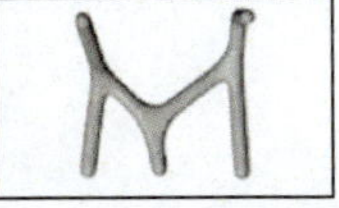

Additionally, frame-like geometries similar to topology optimization, such as the frame of a bus or train, can also be reconstructed. The following sample shows the skin subdivision surface parametrization applied on a train/bus frame.

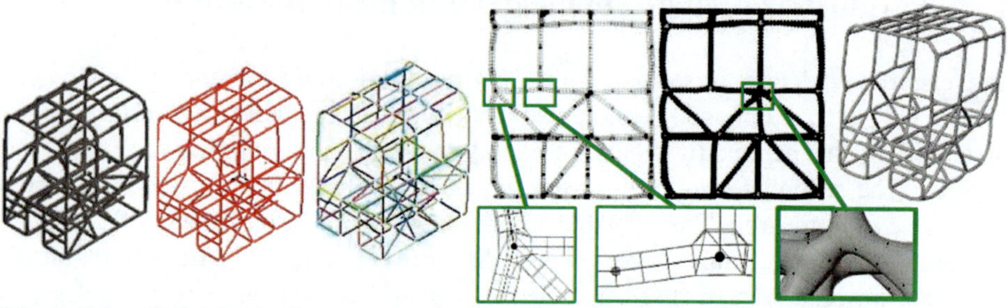

Fig. 5. Skin subdivision surface parametrization on a train/bus frame structure

4.2 Shape Adjustment on Skin Subdivision Surfaces

The parametrization of objects of a non-tube-shaped geometry such as the thin wall samples (2,3,4) can lead to small diameters for the shape parametrization. Due to the flexible underlying concept of skin-based modeling, the geometry can easily be adjusted by moving the skeleton control points to improve the shape coverage. Additionally, the size of the sphere at the corresponding point can be changed so that the overlay is increased. The following figure shows the adjustment of sample three.

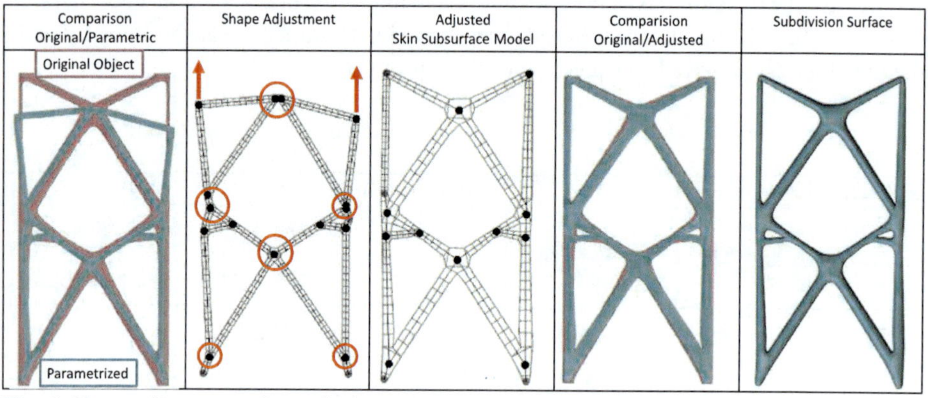

Fig. 6. Shape adjustment of sample three

First, the skeleton points are moved to the location of the orange arrows. Second, the radius of the skeleton points is changed in size. So, by moving two control points and by changing six times the radius, the parametrized skin subsurface model matches the original shape visually.

5 Conclusion

This article presents an automated parameterization of polygon meshes into a parametric representation using subdivision surfaces, especially in topology optimization. For this purpose, we estimate the beam-lines with a skeleton determined by a homotopic thinning method. By additionally using the Euclidean distance transformation, the skeleton lines are transformed into a subdivision surface control grid. This procedure is tested on five topology optimized samples. Parameterization with the thinning approach followed by a distance approximation leads to promising results covering the topology and shape of tubular objects as presented by the first and last example in **Table *1***. If the cross-section is not circular, shape fitting by scaling the control grid surfaces is required. The curvature criterion based on the B-spline fit reduces most skeletal segments to a few beam-lines. While this criterion results in a reasonable line cut for samples one, two, four, and five, sample three must be adjusted, as shown in **Fig. *6***. The manual shape adjustment of the skin subdivision area representation of sample three shows a more flexible and promising parametrization. With a small number of interactions, the shape can be changed until the original representation is covered. For further research investigations, the proposed subdivision skin surface parameterization can be extended to cover rectangle control grids. Additionally, the cutting criteria using line curvature should be extended to additionally covering the normal vector along the skeletal segment. While this approach is limited to organic tube-like objects, the parametrization can also be extended to arbitrary cross-sections. Additionally, despite to frame like objects, the surface skeleton can extend the beam like parametrization to a thin wall like subdivision parametrization.

6 Acknowledgment

This study was funded by the German Federal Ministry of Economic Affairs and Energy (ZF4428401BZ7) in the context of the research network IraSME and ZIM. Responsibility for the content of this publication lies with the authors.

References

1. R. Bénière, G. Subsol, G. Gesquière, F. Le Breton, and W. Puech, 'A comprehensive process of reverse engineering from 3D meshes to CAD models', *Comput.-Aided Des.*, vol. 45, no. 11, pp. 1382–1393, Nov. 2013, doi: 10.1016/j.cad.2013.06.004.
2. V. Vidal, C. Wolf, and F. Dupont, 'Mechanical Mesh Segmentation and Global 3D Shape Extraction', 2014.
3. S. Gauthier, W. Puech, R. Bénière, and G. Subsol, 'Analysis of digitized 3D mesh curvature histograms for reverse engineering', *Comput. Ind.*, vol. 92–93, pp. 67–83, Nov. 2017, doi: 10.1016/j.compind.2017.06.008.

4. A. Ben Makhlouf, B. Louhichi, M. A. Mahjoub, and D. Deneux, 'Reconstruction of a CAD model from the deformed mesh using B-spline surfaces', *Int. J. Comput. Integr. Manuf.*, vol. 32, no. 7, pp. 669–681, Jul. 2019, doi: 10.1080/0951192X.2019.1599442.
5. B. Louhichi, G. N. Abenhaim, and A. S. Tahan, 'CAD/CAE integration: updating the CAD model after a FEM analysis', *Int. J. Adv. Manuf. Technol.*, vol. 76, no. 1, pp. 391–400, Jan. 2015, doi: 10.1007/s00170-014-6248-y.
6. Y. M. Yoely, O. Amir, and I. Hanniel, 'Topology and shape optimization with explicit geometric constraints using a spline-based representation and a fixed grid', *Procedia Manuf.*, vol. 21, pp. 189–196, Jan. 2018, doi: 10.1016/j.promfg.2018.02.110.
7. M. Bremicker, M. Chirehdast, N. Kikuchi, and P. Y. Papalambros, 'Integrated Topology and Shape Optimization in Structural Design∗', *Mech. Struct. Mach.*, vol. 19, no. 4, pp. 551–587, Jan. 1991, doi: 10.1080/08905459108905156.
8. A. Agathos, I. Pratikakis, S. Perantonis, N. Sapidis, and P. Azariadis, '3D Mesh Segmentation Methodologies for CAD applications', *Comput.-Aided Des. Appl.*, vol. 4, no. 6, pp. 827–841, Jan. 2007, doi: 10.1080/16864360.2007.10738515.
9. D. Reniers and A. Telea, 'Part-type Segmentation of Articulated Voxel-Shapes using the Junction Rule', *Comput. Graph. Forum*, vol. 27, no. 7, pp. 1845–1852, 2008, doi: 10.1111/j.1467-8659.2008.01331.x.
10. C. Feng, A. C. Jalba, and A. C. Telea, 'Part-Based Segmentation by Skeleton Cut Space Analysis', in *Mathematical Morphology and Its Applications to Signal and Image Processing*, Cham, 2015, pp. 607–618, doi: 10.1007/978-3-319-18720-4_51.
11. A. Nana, J.-C. Cuillière, and V. Francois, 'Automatic reconstruction of beam structures from 3D topology optimization results', *Comput. Struct.*, vol. 189, pp. 62–82, Sep. 2017, doi: 10.1016/j.compstruc.2017.04.018.
12. T. Stangl and S. Wartzack, 'Feature based interpretation and reconstruction of structural topology optimization results', in *Proceedings of the 20th International Conference on Engineering Design (ICED15)*, Jul. 2015, p. Vol. 6, 235-245.
13. A. Tagliasacchi, T. Delame, M. Spagnuolo, N. Amenta, and A. Telea, '3D Skeletons: A State-of-the-Art Report', *Comput. Graph. Forum*, vol. 35, no. 2, pp. 573–597, May 2016, doi: 10.1111/cgf.12865.
14. M. Denk, K. Rother, and K. Paetzold, 'Multi-Objective Topology Optimization of Heat Conduction and Linear Elastostatic using Weighted Global Criteria Method', in *Proceedings of the 31st Symposium Design for X (DFX2020)*, Bamberg, Sep. 2020, pp. 91–100, doi: 10.35199/dfx2020.10.
15. T. C. Lee, R. L. Kashyap, and C. N. Chu, 'Building Skeleton Models via 3-D Medial Surface Axis Thinning Algorithms', *CVGIP Graph. Models Image Process.*, vol. 56, no. 6, pp. 462–478, Nov. 1994, doi: 10.1006/cgip.1994.1042.
16. P.-S. Tang and K.-H. Chang, 'Integration of topology and shape optimization for design of structural components', *Struct. Multidiscip. Optim.*, vol. 22, no. 1, pp. 65–82, Aug. 2001, doi: 10.1007/PL00013282.
17. H. Blum, *A Transformation for Extracting New Descriptors of Shape*. M.I.T. Press, 1967.
18. K. Siddiqi and S. Pizer, Eds., *Medial Representations: Mathematics, Algorithms and Applications*. Springer Netherlands, 2008.
19. P. Keller, M. Bertram, and H. Hagen, 'Reverse engineering with subdivision surfaces', *Computing*, vol. 79, no. 2, pp. 119–129, Apr. 2007, doi: 10.1007/s00607-006-0191-1.

20. K. Bandara, T. Rüberg, and F. Cirak, 'Shape optimisation with multiresolution subdivision surfaces and immersed finite elements', *Comput. Methods Appl. Mech. Eng.*, vol. 300, pp. 510–539, Mar. 2016, doi: 10.1016/j.cma.2015.11.015.
21. V. Estellers, F. Schmidt, and D. Cremers, 'Robust Fitting of Subdivision Surfaces for Smooth Shape Analysis', in *2018 International Conference on 3D Vision (3DV)*, Sep. 2018, pp. 277–285, doi: 10.1109/3DV.2018.00040.
22. M. Denk, K. Paetzold, and K. Rother, 'Feature line detection of noisy triangulated CSGbased objects using deep learning', in *Proceedings of the 30th Symposium Design for X*, Jesteburg, Germany, Sep. 2019, vol. 30, pp. 239–250, doi: 10.35199/dfx2019.21.
23. P. Wang *et al.*, '3D shape segmentation via shape fully convolutional networks', *Comput. Graph.*, vol. 76, pp. 182–192, Nov. 2018, doi: 10.1016/j.cag.2018.07.011.
24. M. Mortara, G. Patanè, M. Spagnuolo, B. Falcidieno, and J. Rossignac, 'Plumber: a method for a multi-scale decomposition of 3D shapes into tubular primitives and bodies', 2004.
25. H. Xu, M. Dong, and Z. Zhong, 'Directionally Convolutional Networks for 3D Shape Segmentation', in *2017 IEEE International Conference on Computer Vision (ICCV)*, Venice, Oct. 2017, pp. 2717–2726, doi: 10.1109/ICCV.2017.294.
26. A. Tagliasacchi, I. Alhashim, M. Olson, and H. Zhang, 'Mean Curvature Skeletons', *Comput. Graph. Forum*, vol. 31, no. 5, Art. no. 5, 2012, doi: 10.1111/j.1467-8659.2012.03178.x.
27. A. Sobiecki, H. C. Yasan, A. C. Jalba, and A. C. Telea, 'Qualitative Comparison of Contraction-Based Curve Skeletonization Methods', in *Mathematical Morphology and Its Applications to Signal and Image Processing*, 2013, pp. 425–439.
28. J. Mayer and S. Wartzack, 'Ermittlung eines Skelettierungsverfahrens zur Konvertierung von Topologieoptimierungsergebnissen', in *Proceedings of the 31st Symposium Design for X (DFX2020)*, Bamberg, Sep. 2020, pp. 111–120, doi: https://doi.org/10.35199/dfx2020.12.
29. G. Yin, X. Xiao, and F. Cirak, 'Topologically robust CAD model generation for structural optimisation', *Comput. Methods Appl. Mech. Eng.*, vol. 369, p. 113102, Sep. 2020, doi: 10.1016/j.cma.2020.113102.
30. J.-C. Cuillière, V. François, and A. Nana, 'Automatic construction of structural CAD models from 3D topology optimization', *Comput.-Aided Des. Appl.*, vol. 15, no. 1, pp. 107–121, Jan. 2018, doi: 10.1080/16864360.2017.1353726.
31. K. Müller, L. Reusche, and D. Fellner, 'Extended subdivision surfaces: Building a bridge between NURBS and Catmull-Clark surfaces', *ACM Trans. Graph.*, vol. 25, no. 2, pp. 268–292, Apr. 2006, doi: 10.1145/1138450.1138455.
32. P. Min, *binvox*. 2004.
33. F. S. Nooruddin and G. Turk, 'Simplification and Repair of Polygonal Models Using Volumetric Techniques', *IEEE Trans. Vis. Comput. Graph.*, vol. 9, no. 2, Art. no. 2, 2003.
34. P. Dierckx, 'Algorithms for smoothing data with periodic and parametric splines', *Comput. Graph. Image Process.*, vol. 20, no. 2, pp. 171–184, Oct. 1982, doi: 10.1016/0146-664X(82)90043-0.
35. P. Dierckx, *Curve and Surface Fitting with Splines*. USA: Oxford University Press, Inc., 1993.

A Concept Towards Automated Reconstruction of Topology Optimized Structures Using Medial Axis Skeletons

Johannes Mayer[1] and Sandro Wartzack[1]

[1] Engineering Design, Friedrich-Alexander-Universität Erlangen-Nürnberg, Martensstraße 9, 91058 Erlangen, Germany

Abstract. Through continuous topology optimization the design within product development of structural loaded components is enhanced by using maximized lightweight potential. At the same time the product development process is accelerated due to a reduced number of design cycles. Because topology optimization yields an optimized material distribution, the designer is not required anymore to iteratively work out the basical geometric shape. However, optimized material distribution itself neither follows design rules, nor does it provide parametric relations within parts. Instead, a redesign has to be done in order to fulfill manufacturing requirements and to obtain a parametric CAD-model. This takes effort in time and expert knowledge.

For the reason of a consistent development approach under assistance of computer aided methods, a concept is presented, that is demanded to solve the task of redesigning topology optimization results in parametric CAD-models in a semi-automated manner. To achieve this, medial axis skeletons are used for analyzing the geometric data firstly, and, in a second step, for rebuilding CAD-models. They are taken as references for common CAD-functionality whereby a parametric and history-based modeling structure is generated. Though open challenges exist in further automation and consideration of manufacturing techniques, this represents a basical approach on how to use the medial axis for reconstruction.

Keywords: Topology Optimization, Medial Axis Skeleton, CAD.

1 Redesign of topology optimized structures

Having to define the geometry of a structural component during design stage in the product development process (PDP), design engineers are free to use structural optimization tools like topology optimization (TO). In the common approach with solid isotropic material penalization (SIMP), this finite element based iterative method determines a structural optimized material distribution for a given loading situation [1, 2, 3]. Saving repeated design cycles, an initially overall design space this way directly results in a design proposal [1]. In addition to time savings in the PDP, this allows the construction of load-specific-designed, functional lightweight structures [1, 2, 3].

S. Pfingstl et al. (eds.), *Proceedings of the Munich Symposium on Lightweight Design 2020*, https://doi.org/10.1007/978-3-662-63143-0_3

However, because the design proposal is derived from a discretized mesh-model, it is far from being a finalized, fully-designed part [1, 3]. Instead the design proposal usually consists of a triangulated surface representation (STL-file). It neither meets manufacturing requirements nor parametric or feature-based model representation. Even the surface representation is typically characterized by an unsmooth boundary, where in a designer's interpretation, for example, a flat face would be (see Fig. 1) [1].

Fig. 1. Model data at significant design stages (design space from [4]).

While manufacturability is a must, for a productive product development it is also necessary to have access to a part model that is easy to modify. This is reached through a parametric, feature-based and -if desired- history-based modeling structure. Therefore the design proposal has to be specifically redesigned in a new CAD-model corresponding the demands of the PDP. [5]

For usual, the task of redesign is done manually. Despite its requirement in expertise on engineering design, this is a time consuming process, especially for complex shapes [5].

More automated software tools allow for an assisted redesign, for example in remodeling the surface with NURBS-patches. While this can be time saving, its result is still a surface model with very restricted, or no parametric relations nor design history. [5]

Another approach is the skeletonization-based abstraction of geometry followed by partially automated redesign with common CAD-features. In [5] a method is presented, that uses a skeletonization algorithm to convert the topology optimization result in the decomposed shape of simple curves. These curve lines afterwards are used as references for sweeping cross sections. By means of this principle, the result will be a referenced, still modifiable, volumetric part. Within this method there is user input and interaction necessary, e.g. to refine the skeleton with curve splines. Further user interaction is required for the sweeping feature or for the redesign of connection elements. The curve skeleton abstraction is beyond that not generally suited for the abstraction of topology optimized geometry. In [6] the author concluded challenges to abstract surface-like geometry with curve skeletonization algorithms. [5]

Nevertheless, the approach of skeletonization-based topology reconstruction is the method that meets the PDP's demands in parametric model-structure. The advantages on the one hand, lie in the ability to abstract the principal shape of a geometry with skeletons [7], on the other hand, in the similar design practice to CAD-design through the use of common CAD-functionality [5].

For the reason of a more assisted and faster development process subsequent to the application of topology optimization, the task is to develop an approach for automated topology reconstruction using skeletonization methods, and meeting the demands of the PDP, as well as manufacturing requirements of basic manufacturing technology. In the following sections, insights into this task are given, and a possible semi-automatic concept is presented.

2 An approach towards a more automated reconstruction of topology optimization results

2.1 Analysis

In 1969 Blum gave a definition of skeletons for 2D-objects: it is the locus of centers of maximally inscribed circles [8]. Blum's definition found use in further development in the field of computer graphics and subsequent skeletonization algorithms [9]. It was extended into the third dimension and used for surface reconstruction [10]. Here, the set of centers and their respective radii together are the so-called medial axis transform [9, 10]. While computing the exact medial axis transform is difficult, a simplified approximation is less computational expensive and more useful [10].

What exactly makes a skeleton, is up to definition according to [7]. Thus, to this day, there are several more skeletonization methods to be distinguished [7], making the curve skeleton used in [5] one amongst many. While this one falls under the terms of curve skeletons (consisting of one-dimensional curves), the so-called surface skeletons also contain surface elements. The medial axis skeletons belong to this latter category. [7, 10]

In the presented approach to restructuring topology optimized results, two main steps are declared. The first one is the analysis-step, where the geometric shape has to be abstracted. The second step is the synthesis-step, which describes the redesign on the basis of the prior decomposition (see Fig. 2).

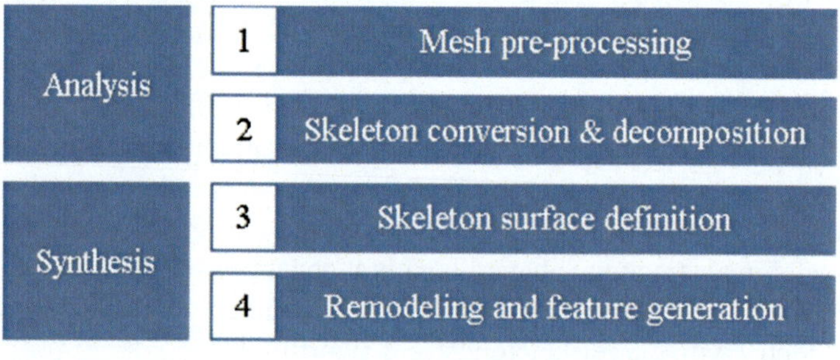

Fig. 2. Basic steps in the presented approach of topology restructuring

Initially, it is aimed for a conversion of the design proposal into information suited for reconstruction. This means, that structural topology optimized results, as various as they can be, are able to be captured by a defined scheme of decomposition.

To achieve this, medial axis skeletons are used. With them, the triangulated design proposal can be skeletonized. As soon as the optimization is done, a triangulated surface of the design proposal is extracted. This needs to be pre-processed to obtain a watertight STL-file without edge interferences or overlapping facets (Fig. 2, 1). If the exclusively subtracted material distribution is triangulated alike, this geometry can be skeletonized, too. In each case there are two skeletal forms obtained – one inside ("skeleton") and one outside ("skeleton complement") the geometry [7, 10].

Due to the calculation of maximal inscribed balls in the medial axis transform, the skeleton is strongly related to the geometric shape by the additional information of each of these balls' radius. In terms of topology reconstruction from a design proposal, this information is to be seen as local cross section thickness. Because the STL-file representing the design proposal is a discrete data set (set of xyz-coordinates plus connectivity), the skeleton with the cross section information again is a similar discrete data set. Compared to principally other skeletonization algorithms, the local thickness information is advantageous [11], as it is assumed, that the topology-restruct-method has to gather as much information as possible for a viable decomposition.

For decomposition, it's furthermore possible to separate the geometry in distinctive classes. One option is to differ effective surfaces and accompanying geometry areas (ES), as well as the structural connection between those, the lead support structure (LSS) according to [12]. The redesign task then consists of rebuilding ES and LSS. This separation scheme is thought to fit to topology optimized geometry in terms of regular and irregular geometric shape. As initially mentioned, topology optimized material distribution typically doesn't show geometric primitives [1, 2, 3, 5]. However, in the product design of structural components, the interface areas to neighboring parts are usually predefined. Accordingly, these sub-areas of the geometry are excluded from the optimization beforehand (frozen area) [1, 2, 3]. This way their principally regular geometric shapes can be found in the design proposal as geometric primitives. By dividing in ES and LSS, it's possible to capture the geometric primitives as effective surface accompanying geometry and all of the optimized material distribution as LSS.

Besides the ES which should be redesigned through geometric primitives, the LSS has to be provided with another reconstruction scheme. The medial axis skeleton is capable of abstracting a surface representation from the LSS, but can't be reconstructed directly into construction geometry. A thinkable option is to approximate the boundary of the medial axis skeleton to define one or more distinctive surfaces. Together with the information about cross section thickness and the geometric primitives at interface areas, this would be the decomposition of topology optimized structures (see Fig. 2, 2). The decomposition is the result of the analysis-step.

2.2 Synthesis

After firstly analyzing the design proposal, the second main step consists of rebuilding the structural geometry in order to gain a parametric model for further use in the PDP. Since the desired method is expected to suit general redesign tasks to fit in the PDP, the conceptual synthesis-step preferably has to be generally applicable.

In the presented approach, the decomposition is used to reconstruct the lead support structure followed by the predefined effective surface geometry. The ES accompanying geometry is adopted from the earlier phase of the specification of frozen area in TO (see Fig. 2, 4). Though this transition of geometry from the design space stage into the stage of redesign could be done in a small amount of time, it's suited for automation, since it's always a similar, repetitive task. If there is no data to frozen areas in the TO available and the design proposal itself is the only input into the topology reconstruction, the automated redesign of ES immediately gets more challenging because of the now necessary identification of those ES. An option to do so, would be, to use an algorithm for primitive detection [13]. Suited to identify geometric primitives in point clouds [13], it could be adapted for the automated reconstruction of ES.

Redesigning the LSS in the synthesis-step is done on basis of the decompositions skeleton boundary and cross section radius. The skeletal boundary defines one ore more skeletal surfaces which approximate the design proposal's geometry. These surfaces have to be constructed in the CAD-environment, which again is a task to be automated (see Fig. 2, 3). Challenges herein lie in the identification of an explicit surface boundary and distinctive surface patches, as well as the guaranty of generally application. How to best redesign CAD-geometry from this kind of data has to be investigated. One possible way is to shape each surface in the surface-set in a way that it fits the medial axis skeleton, and offset it afterwards to both sides with an extrude-like operation at the same time to yield a volumetric body (see Fig. 2, 4). The offset distance is taken from the information about the cross section radius. Finally some chamfers and rounding are inserted where necessary. At the end of the synthesis-step, a parametric CAD-model is available.

3 Redesign of a turbine engine bracket with a more automated approach to reconstruction of topology optimization results

The presented approach should be demonstrated with the practical example of a turbine engine bracket. The design space geometry (see Fig. 1) originates from a public idea challenge, where the manufacturer asked for the best additive-manufacturing-design in terms of functional and lightweight aspects. Within the list of entries, it's evident that the TO was widely used to determine a principal geometric shape. [4]
After a corresponding topology optimization, the analysis-step is entered. Herein, the design proposal is prepared for the skeletal conversion through cleaning and isotropic re-meshing, resulting in an input-surface-mesh (see Fig. 3, 1). Then the curve and surface skeletons are computed (Fig. 3, 2a and 2b). The surface skeleton is computed with a voronoi-based method as a discrete scheme of the medial axis transform, related to the algorithm described in [10]. Within the computation, the input-surface-mesh is slightly smoothed. The connectivity of skeletal nodes origins by the connectivity of the input-surface-mesh [9]. As described above, the medial axis transform is the set of centers of maximally inscribed balls and their respective radii. Plotted altogether, the balls form the so-called union of balls, which is a surface-approximating geometry (see Fig. 3, 2c) [10].

As soon as the medial axis transform is available, the boundary of the skeleton is computed (see Fig. 3, 2d). The algorithm detects boundary points of the medial axis skeleton, but does not differ between possible sections of the medial surface they could belong to. While this is up to user intervention and requires his interpretation for now, it is an opportunity for enhancement of this approach.

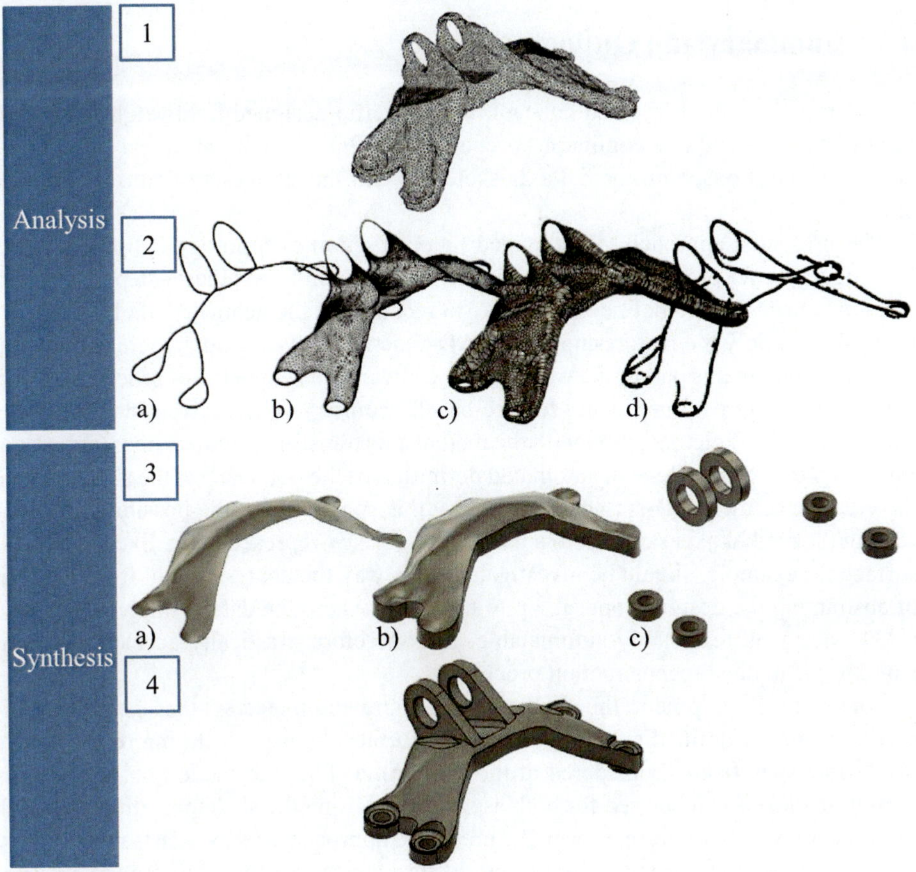

Fig. 3. Data representation at basical steps in the presented approach towards a more automated reconstruction of topology optimized structures. A turbine engine bracket acts as demonstrator: 1. pre-processed surface mesh; 2. decomposition data; 3. CAD-reconstruction-data; 4. resulting CAD-geometry

The first steps in the synthesis start with the automated loading of the decomposition data into the CAD-system. Then the skeletal boundary is turned into a bounding spline and a skeletal surface created by this spline (Fig. 3, 3a). During the creation, the surface is oriented to contain relevant sections of the curve skeleton. This is done to inherently imitate the basical shape of the design proposal. Afterwards the skeletal surface is offsetted in the left and the right perpendicular direction and connected through a loft-operation. Next, the recently created volume body is thickened in the planar direction (see Fig. 3, 3b). Now, the ES are reconstructed (see Fig. 3, 3c). In the final moves, the geometry is manipulated with boolean operations to combine the geometry on the one hand, and to subtract sections, which cross the design space, on the other hand. Lastly, some chamfers and roundings are inserted, and the CAD-model reconstruction is completed (see Fig. 3, 4). Due to the use of built-in functionality in CAD, a design history is obtained, opening up the freedom to subsequent changes at any place.

4 Summary and Outlook

By the means of the presented concept, load-specific designed lightweight structures can be designed within a continuous process in the PDP. Important aspects are the geometry analysis based on medial axis skeletons and the synthesis of construction geometry.

The presented approach to automated reconstruction of topology optimized structures does not yet provide a fully automated solution. In the synthesis-step, user intervention is necessary to define the surface, to reconstruct the geometry of effective surfaces and to apply the feature-input for CAD-functionality. Although there is potential for automation in this part of the workflow, the advantage is the kind of procedure itself, with its resulting parametric and feature-based geometry. The analysis-step is widely automated. The skeletonization with the medial axis transform requires no intervention. Enhancements can be done in automated definition of the boundary of the skeleton, and the creation of the CAD-surface bounded by this spline. Here, the possibility of converting the medial axis skeleton to a parametric surface-representation, like a NURBS-surface for example, should be investigated. This way the analysis, right from the start of abstracting the design proposal, up to the point, where the data is imported into the CAD-system, is reasonably automatable, and therefore drastically accelerated compared to the manual reconstruction process.

For future development, the challenge of a more automated synthesis could be addressed with pre-defined geometry elements. Further the manufacturing requirements need to be specifically considered at the same time. The information on local cross-section-thickness can be used for both aspects, exploiting the skeletonization approach to a higher level. In its current form, the presented approach already assists the topology redesign in interpreting the design proposal through the various decomposition data. Adjoining, the presented synthesis-procedure is a scheme, a design engineer can consult, when having to redesign topology optimized structures.

5 Acknowledgements

This report is the scientific result of a research project undertaken by the DFG (German Research Foundation) and the Chair of Engineering Design at the University of Erlangen-Nürnberg under the direction of Prof. Dr.-Ing. Sandro Wartzack.

The authors thank the German Research Foundation for funding the project WA 2913/36-1 “TopoRestruct – Converting topology optimization results into a design geometry, which meets the requirements for manufacturability, functionality and mechanical stress in the product development process”.

References

1. Harzheim, L.: Strukturoptimierung. Grundlagen und Anwendungen. 2nd edn. Europa Lehrmittel, Haan-Gruiten (2014).
2. Bendsøe, M. P., Sigmund, O.: Topology Optimization – Theory, Methods and Applications. Springer, Berlin 2004.
3. Schumacher, A.: Optimierung mechanischer Strukturen. Grundlagen und industrielle Anwendungen. 2nd edn. Springer, Berlin 2013.
4. GrabCad Homepage, https://grabcad.com/challenges/ge-jet-engine-bracket-challenge, last accessed 2020/10/14.
5. Stangl, T., Wartzack, S.: Feature Based Interpretation and Reconstruction of Structural Topology Optimization Results. In: Proceedings of the 20th International Conference on Engineering Design (ICED15), vol. 6: Design Methods and Tools, Milan, Italy, 27.-30.07.2015.
6. Kresslein, J., et. al.: Automated cross-sectional shape recovery of 3D branching structures from point cloud. In: Journal of Computational Design and Engineering, vol. 5, pp. 368–378 (2017).
7. Tagliasacchi, A., et al.: 3D Skeletons: A State-of-the-Art Report. In: Madeira, J.; Patow, G.; Romão, T. (eds.): Computer Graphics Forum 2016, vol. 35 nr. 2, pp. 573–597. Wiley & Sons Ltd.
8. Blum, H.: A Transformation for Extracting New Descriptors of Shape. In: Wathen-Dunn, Weiant (ed.): Models for the Perception of Speech and Visual Form. MIT Press, 1967. pp. 362–380.
9. Siddiqi, K., Pizer, S. M.: Medial Representations. Mathematics, Algorithms and Applications.Springer, New York, NY, USA 2008.
10. Amenta, N. Choi, S., Kolluri, R. K.: The Power Crust. In: Proceedings of the 6th ACM Symposium on Solid Modeling and Applications 2001, pp. 249–266. ACM, New York, NY, USA (2001).
11. Mayer, J., Wartzack, S.: Ermittlung eines Skelettierungsverfahrens zur Konvertierung von Topologieoptimierungsergebnissen. In: Krause, D., Paetzold, K., Wartzack, S. (eds.) Proceedings of the 31st Symposium Design for X (DFX2020) 2020, vol. 31, pp. 111–120. (2020)
12. Matthiesen, S.: Ein Beitrag zur Basisdefinition des Elementmodells "Wirkflächenpaare & Leitstützstrukturen" zum Zusammenhang von Funktion und Gestalt technischer Systeme. Dissertation, Universität Karlsruhe, 2002.
13. Schnabel, R., Wahl, R., Klein, R.: Efficient RANSAC for Point-Cloud Shape Detection. Computer Graphics Forum 2007, vol. 26 nr. 2, pp. 214–226. Blackwell Publishing Ltd.

Design and Optimization of Ultra-Stable Fine-Pointing Structures for the CHIME Instrument

Markus J. Geiss, Peter Buschkamp, Herbert Loew, Bernhard Sang, and Stephan Rapp

OHB System AG, Manfred-Fuchs-Str. 1, 82234 Wessling/Oberpfaffenhofen
Markus.Geiss@ohb.de

Abstract. CHIME (Copernicus Hyperspectral Imaging Mission for the Environment) is part of ESA's Sentinel Expansion Program for which OHB System AG in Oberpfaffenhofen, Germany has been selected as the instrument prime contractor. In this paper we present the development of the ultra-stable stiffness architecture for the instrument structure. Backed by a trade study of different structural concepts with a focus on the opto-mechanical decoupling, assembly integration and test accessibility, and the demanding optical performance criteria, we derive a robust solution: The main stiffness element is a central monolithic CFRP (Carbon Fiber Reinforced Plastic) torus, on which several sub-elements are mounted for precise positioning of the highly-sensitive optical elements, creating a modular 3D optical bench. We demonstrate the use of topology optimization for design concept finding considering mass and performance criteria. The paper concludes with the results from a breadboard test campaign and an outlook on the next development steps.

Keywords: Opto-mechanics, monolithic CFRP, Space Telescope, Optical Instrument, Topology Optimization, Breadboard Testing

1 Introduction

The CHIME mission is ESA's future hyperspectral imaging mission in a sun synchronous LEO orbit continuously mapping the earth surface at 30m ground resolution in the full Visible and Near Infrared wavelength range from 400 to 2500nm at 10nm spectral resolution. A multi-satellite constellation at an altitude of 632km ensures timely revisit of ground points. Thales Alenia Space, France is the mission prime contractor with OHB System AG, Germany being instrument prime contractor. OHB's know-how and experience from the German Space Agency's hyperspectral Satellite EnMAP[4] program have been used throughout the phases A and B1 to develop the CHIME instrument and mission concepts, securing the best possible solution while managing the associated risks at reasonable costs. All elements of the optical and electro-optical imaging chain are held by a single stiffness-providing element, a central monolithic CFRP torus, creating an ultra-stable stiffness architecture for the instrument's structure. This

S. Pfingstl et al. (eds.), *Proceedings of the Munich Symposium on Lightweight Design 2020*, https://doi.org/10.1007/978-3-662-63143-0_4

creates a design with robust margin on the demanding stability requirements of this mission in Line-of-Sight stability, pointing accuracy and spatial and spectral co-registration of the recorded data. This paper gives an overview over a comprehensive design trade study activity which was conducted in the phase A/B1 system study to obtain the conceptual layout of the structural sub-system. Furthermore, the detailed design process of the spectrometer sub-structure via topology optimization is discussed. Finally, results from a structural hardware test are presented and an outlook into the next project phase is given.

2 Design Trade-Offs

The structural architecture of the CHIME instrument is driven by the following key parameters:

- Ensure structural integrity against environmental loads (launch environment, thermal loads) whilst minimizing mass.
- Consequent frequency decoupling between spacecraft, instrument primary structure and instrument sub-systems to minimize response amplitudes caused by micro/launch vibrations and thus limit related design loads.
- Minimize global thermo-elastic deformations in space to comply with allocated optical performance budget.
- Provide adequate iso-static interface (IF) decoupling to minimize the impact on optical performance due to thermo-elastic distortion and IF tolerances. This also ensures good test-ability on sub-system level.
- Provide sufficient free space and unobstructed view between sub-systems to allow for Assembly, Integration and Test (AIT) handling and alignment.

To achieve those, partially contradicting, criteria in the best possible manner, we performed a comprehensive trade study. The general process flow of it is summarized in Figure 1. It should be noted that the Finite Element (FE) models are shown only for illustration purposes to represent the general idea of a closed versus a more open design concept using different materials and/or material combinations.

2.1 Honeycomb Panel Design

One option resulting from the material trade is a compound of honeycomb composite panels which constitute a 3D optical bench in a non-classical sense. The FE model of such a design concept is shown in Figure 1 at the top. Depending on the material combination used, this panel concept has the potential to exhibit low thermal expansion which in turn leads to relaxed thermal control requirements. In a mixed material situation where CFRP is used for structural elements and glass or aluminium is for optical elements, adequate opto-mechanical decoupling is needed in-between the different elements and sub-systems of the instrument. The FE analysis of the panel design in the traded material combination showed that CFRP sandwich panels with aluminum core perform best with respect to

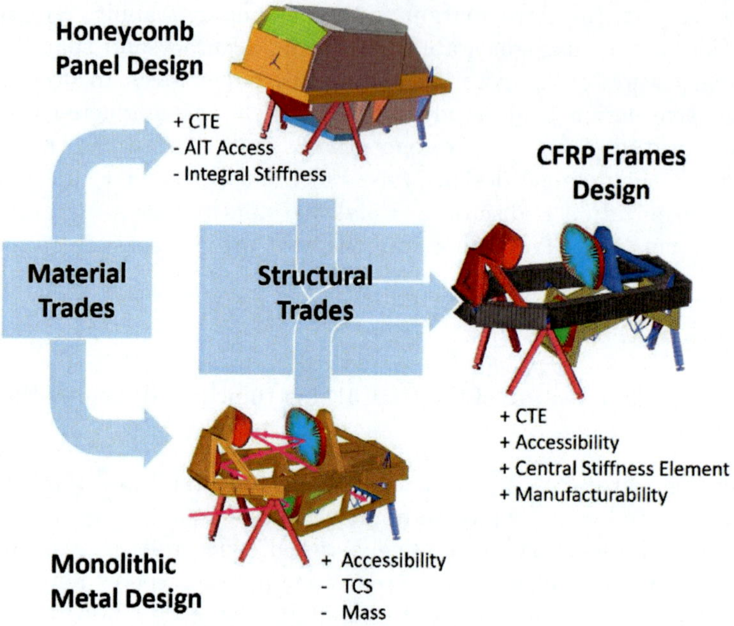

Fig. 1. Process flow of material trades and concept trades yielding the final CHIME instrument structural concept. The chief array of the optical path is shown in magenta.

the given stiffness requirements. Their beneficial stiffness to weight ratio makes honeycomb sandwich panels typically the preferred choice for classical 2D optical benches as for example used in the Meteosat Third Generation instruments [2]. However and most importantly, the 3D arrangement of the CHIME instrument panel design obtains its stiffness only on an ensemble level, i.e. only if all panels work together and maintain their relative alignment, is the structure stiff enough to provide the necessary functionality. To realize this, the necessary design, manufacturing and AIT efforts become prohibitive and thus extremely costly if the CHIME instrument was realized by a pure honeycomb panel design.

2.2 Monolithic Metal Design

In contrast to a differential design composed of several sandwich panels, a monolithic metal design concept is investigated next. One possible design of such a concept is shown in Figure 1 at the bottom (along with the chief array of the optical path in magenta). Since stability is identified as the key decision driver, it is clear that a monolithic metallic design shows a high thermal deformation creating the need for a highly sophisticated Thermal Control System (TCS). However, the number of structural IFs is minimal and no adhesives are used which increase the predictability of such a monolithic metal structure. In addition, the AIT accessibility is provided by definition. On the other hand, the mass required to achieve adequate stiffness is larger than for the honeycomb panel design concept. Due to the large size of the CHIME instrument, subtractive manufacturing of a corresponding monolithic metal structure yields significant schedule and cost risks and should therefore be avoided.

2.3 CFRP Frame Design

Hinging upon the advantages of both the panel design and the monolithic design, a design made from CFRP frame structures is investigated next. This merged concept is shown in Figure 1 on the right. The conceptual idea is that the instrument is built around a central torus-like hollow CFRP frame structure (made from monolithic CFRP composite with no honeycomb) which is light-weight and provides the necessary opto-mechanical stiffness through a single element. The overall instrument mass is about 450kg. Having one central stiffness providing element increases the predictability of the entire stiffness architecture and thus greatly improves the robustness of the instrument design. At the same time it provides advantageous thermo-elastic stability and maximum access for AIT activities through an open frame concept. The rough dimensions of the instrument outline are 2.8m by 1.2m by 1.4m.
Due to the generally very low global thermal expansion of CFRP, the proposed design concept is rather insensitive to any thermal environment temperature fluctuation in space. Therefore, the degradation of optical performance due to thermo-elastic effects is negligible. This simplifies the entire instrument concept as it does not require a highly sophisticated TCS, i.e. no need for high operational temperature stability of the instrument.
The proposed structural architecture follows a consequent IF and frequency de-coupling approach with the first eigenfrequency of the hard-mounted instrument at 53Hz. Thanks to CFRP as the main material, stiff and light-weight primary and secondary structures with excellent thermo-elastic performance are obtained. The torus-like hollow CFRP frame structure mounts directly onto the platform while all electro-optical and opto-mechanical sub-systems as well as secondary structures connect to this central stiffness element. The CFRP bench performance and initial design has been optimized in several FEM-driven design iterations and the general manufacturability is confirmed by a predevelopment activity (see Section 4 for more details).
The CFRP main frame carries the telescope unit which consists of 4 glass mirrors. All four mirrors connect isostatically to the CFRP main frame via dedicated CFRP sub-frames or directly via kinematic mounts. Towards the rear and bottom of the instrument, a spectrometer unit support structure is isostatically mounted to the main frame. A detailed design approach for this Spectrometer Mounting Structure (SMS) is discussed in Section 3.
One major drawback of CFRP is the moisture release and thus shrinkage of the structure in vacuum and potential contamination of the optical components. The shrinkage and therefore the caused misalignment of the optics has been quantified by a dedicated hygro-elastic analysis. The analysis confirms that the hygro-elastic distortion of the proposed structural architecture can almost entirely be compensated by various means (e.g., refocusing mechanism). In terms

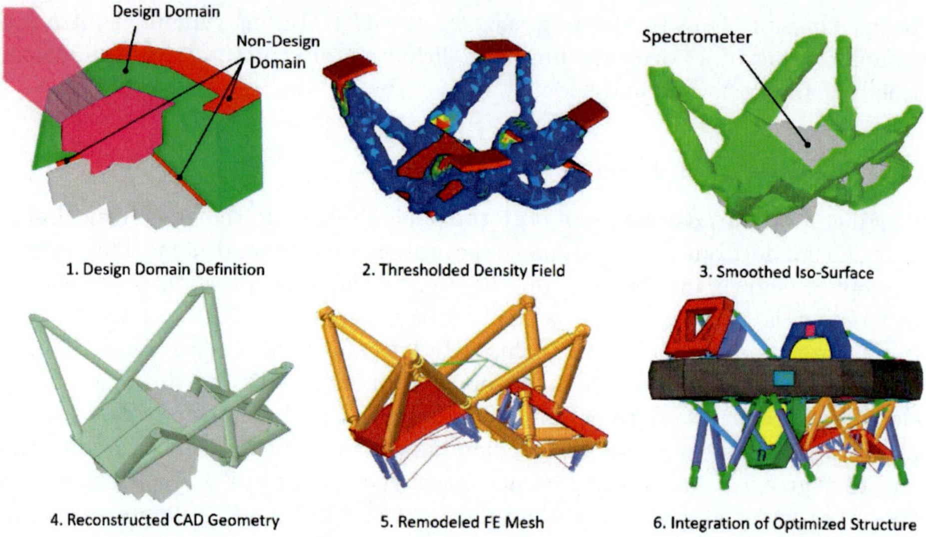

Fig. 2. Topology optimization process flow from the definition of the design domain for the SMS to the integration of the derived optimal structure into the CHIME instrument.

of potential contamination, an analytical assessment of the water contribution and primary deposition areas in vacuum has been performed, to confirm the suitability of the proposed CFRP material.

3 Spectrometer Mounting Structure Design and Optimization

To highlight the computational design approach used to identify the CHIME instrument design concept, the design and optimization process of the spectrometer structure is discussed next. The goal is to design a SMS which connects the spectrometer unit to the main CFRP frame.

To explore the design space of the SMS, topology optimization[3] is employed. The geometrical design domain along with non-design domains (for optical path, calibration mechanism, kinematic mounts) is defined as the starting point of the optimization process flow. Furthermore, the mechanical boundary conditions and load introduction zones are being defined. The optimization problem is formulated as a mass minimization problem subject to displacement constraints on the spectrometer accounting for the extreme stability requirements (sub-micron, sub-arcsec) on the SMS. A worst-case displacement envelope is being used within the optimization considering three different gravity-release load cases. The obtained density distribution, thresholded at 10%, is shown in Figure 2 at the top. Using this thresholded density, a smoothed iso-contour is extracted in a post processing step. Using density-based topology optimization, a non-intuitive, design concept for the quite complex SMS was obtained in a systematic way. The

final solution is comprised of a network of trusses which connects four IF points at the top with two tilted IF planes holding the spectrometer at the bottom. At the same time it is ensured that the geometric envelops of the calibration unit and the optical path are not violated. It can also be seen that the complex final structure obtained by topology optimization can be approximated quite well by one-dimensional bars using CAD. Based on the reconstructed CAD geometry, a simplified beam model is created and re-analyzed to verify the stiffness of the SMS. Using computational design optimization, a non-intuitive design for the SMS comprised of uniaxial struts is identified which fulfills the performance requirements at minimum mass.
Overall, a design solution for the SMS is identified which allows the spectrometer kinematic mounts to all have equal length. Additionally, having the mounting frame strictly above the spectrometer allows for unrestricted access to the focal plane assemblies. This greatly simplifies future AIT efforts.

4 CFRP Frame Hardware Test

To verify the manufacturability and the achievable stiffness performance of the main CFRP frame, a predevelopment activity was performed during the phase B1. In the frame of this activity, the frame design was further detailed in terms of internal stiffening, load introduction zones, wall thickness and cross section geometry. In this respect, OHB builds upon extensive CFRP frame heritage from SUNRISE Solar Observatory[1] and the LIDAR Instrument ALADIN[5]. Finally, a section of the main CFRP frame was manufactured as a breadboard and used for static load testing as shown in Figure 3.

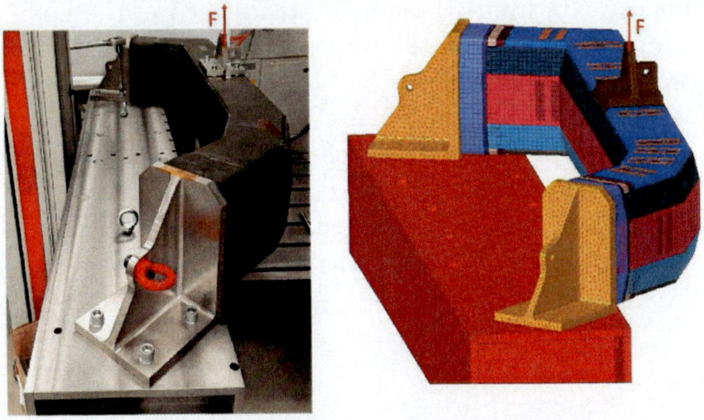

Fig. 3. Breadboard of the main CFRP frame (left) and the corresponding FE model used for static load testing.

To mature the technology readiness of the structure sub-system, static load tests were performed. In a first step, the verification of the technology maturity was done via a specimen static test program representing the critical splice. This CFRP sample program was used to quantify the effect of different manufacturing techniques on the mechanical properties of the material. In addition, the manufactured main frame breadboard is used for static strength testing under different load cases (out-of-plane tensile test and in-plane tensile/shear test). Overall, a good match (less than 30% relative difference) between numerical predictions and physical measurements is obtained for the performed static test cases. To eliminate any settling effects and verify the repeatability of the physical experiments, multiple load and unloading cycles were performed. The results of this predevelopment activity do not only demonstrate the technology readiness in terms of manufacturability but also underline the superior performance of the proposed CFRP frame concept in terms of expected structural behavior.

5 Conclusions and Outlook

This paper summarizes the design and optimization tasks performed in the phase A/B1 system study for the CHIME hyperspectral instrument yielding a fully compliant instrument structure concept. Due to highly challenging performance requirements, a non-classical design approach was taken where the structural architecture (FEM) was optimized and matured first, before only at the very end a mechanical model (CAD) was derived from it. Several different structural architecture concepts were explored until the final, CFRP frame solution was obtained. This highly modular concept exhibits superior thermo-elastic behavior, maximum AIT accessibility and easy manufacturability via a single central stiffness element. All optical elements are mounted iso-statistically onto this central CFRP frame via CFRP sub-frames. Density-based topology optimization was used to derive a truss structure for mounting the spectrometer unit to the instrument main frame. Finally, a breadboard of the main CFRP frame was manufactured and tested in the course of a predevelopment activity. In the subsequent project phases the focus will lay on detailing the proposed stiffness architecture as a whole. While the manufacturability of the main CFRP frame has been demonstrated, all sub-frames as well as the joining concepts need further detailing. Using the already existing breadboard of the main CFRP frame, thermo-elastic and hygro-elastic measurements are planned for numerical model correlation and extrapolation of the frame behavior to the entire CHIME instrument.

References

1. Bittner, H., Erdmann, M., Haberler, P., Härtel, K., Barthol, P. and Curdt, W., Baseline design of the SUNRISE Telescope, Proc. SPIE 5489, Ground-based Telescopes, 2004;
2. Durand, Y., Hallibert, P., Wilson, M., Lekouara, M., Grabarnik, S., Aminou, D., Blythe, P., Napierala, B., Canaud, J., Pigouche, O., Ouaknine, J., and Verez, B., "The flexible combined imager onboard MTG: from design to calibration", Proc. SPIE 9639, Sensors, Systems, and Next-Generation Satellites XIX, 2015;

3. Geiss, M. J., Level-Set-XFEM-Density Topology Optimization of Active Structures: Methods and Applications, Dissertation, University of Colorado at Boulder, 2018;
4. Hofer, S., Förster, K., Sang, B., Stuffler, T., Penne, B., Kassebom, M., Kaufmann, H., Müller, A., Chlebek, Ch.: EnMap Hyperspectral Mission - The Space Segment, Hyperspectral Workshop 2010 Frascati, Italy, 2010;
5. Kaiser, C., Widani, C., Härtel, K., Haberler, P., Lecrenier, O., Buvat, D., and Labruyere, G., Development and verification of a high-performance CFRP structure for the space-borne lidar instrument ALADIN, Proc. SPIE 10567, International Conference on Space Optics — ICSO 2006;

Lightweight engineering design of nonlinear dynamic systems with gradient-based structural design optimization

Erich Wehrle[0000−0002−5760−4468] and Veit Gufler[0000−0002−3349−2406]

Free University of Bozen-Bolzano, Universitätsplatz 1, 39100 Bozen, South Tyrol, Italy
{Erich.Wehrle, Veit.Gufler}@unibz.it

Abstract. Reducing the weight of a system leads to lower forces being exerted, which in turn allows for lower requirements and an even lighter system. This "virtuous circle of lightweight engineering design" can especially be present when designing dynamic systems. Design optimization is a tool to enable and exploit this favorable phenomenon. This work introduces a unified approach to reap the benefits of optimally designed lightweight systems in structural dynamics and multibody dynamics. An efficient gradient-based optimization framework has been implemented and this is explained and demonstrated. The centerpiece of this optimization methodology is the design sensitivity analysis applied to the time integration with a nonlinear solver. A semi-analytical approach is chosen to balance computational effort and implementation effort, where the sensitivities are derived via direct differentiation with numerical differences for the sensitivities of the system parameters. Nomenclature is introduced to simplify these equations for a more lucid description showing the intrinsic equivalence of the solving routines of structural dynamics and multibody dynamics. The method is shown on the practical example for the optimal design of a hydraulic engineering mechanism.

Keywords: Lightweight engineering design · Design optimization · Sensitivity analysis · Structural dynamics · Multibody dynamics

Nomenclature

∇	sensitivities w.r.t. x (total derivative)	F_c	right-hand side of acceleration constraints
∂	partial derivatives w.r.t. x	F_{ext}	external force
$\frac{\partial(\cdot)}{\partial(\cdot)}$	partial derivative	F_{pseudo}	pseudo load
$(\cdot)$	scalar	F_R	residual force
$\underline{(\cdot)}$	vector	F_v	quadratic velocity term
$\underline{\underline{(\cdot)}}$	matrix	d	damping
$\underline{\underline{\underline{(\cdot)}}}$	3D matrix	$\underline{\underline{\underline{\underline{e}}}}$	4D identity matrix
$\underline{\underline{\underline{\underline{(\cdot)}}}}$	4D matrix	i	time step index
		J	Jacobian w.r.t. q
		$\dot{\text{J}}$	Jacobian w.r.t. $\dot{q}$

S. Pfingstl et al. (eds.), *Proceedings of the Munich Symposium on Lightweight Design 2020*, https://doi.org/10.1007/978-3-662-63143-0_5

$\ddot{\mathrm{J}}$	Jacobian w.r.t. $\ddot{q}$	$\dot{q}$	velocity
$\overset{\lambda}{\mathrm{J}}$	Jacobian w.r.t. λ	$\dot{q}_{\mathrm{pred}}$	predicted velocity
$\overset{\nabla}{\mathrm{J}}$	Jacobian w.r.t. ∇q	$\ddot{q}$	acceleration
$\overset{\dot{\nabla}}{\mathrm{J}}$	Jacobian w.r.t. $\nabla\dot{q}$	t	time
$\overset{\ddot{\nabla}}{\mathrm{J}}$	Jacobian w.r.t. $\nabla\ddot{q}$	x	design variables
$\overset{\nabla\lambda}{\mathrm{J}}$	Jacobian w.r.t. $\nabla\lambda$	Δ	finite change
k	stiffness	λ	Lagrangian multipliers
m	mass	Φ	kinematic constraints
q	position	β, γ	integration constants
q_{pred}	predicted position		

1 The Virtuous Circle of Lightweight Engineering Design

The lightweight engineering design philosophy is exemplified by the Virtuous Circle of Lightweight Engineering Design. With less structural mass, the structural requirements, motorization requirements or both are reduced and therefore the structural mass can in turn be reduced again. This can continue until some theoretical minimum is reached. The virtuous circle design philosophy is magnified when looking at dynamic machines. This stands in stark contrast to the vicious circle of excess structural weight. Here extra structural mass increases the structural or motorization requirements, which in turn results in more weight, ending in a concept which no longer is able to fulfill design requirements..

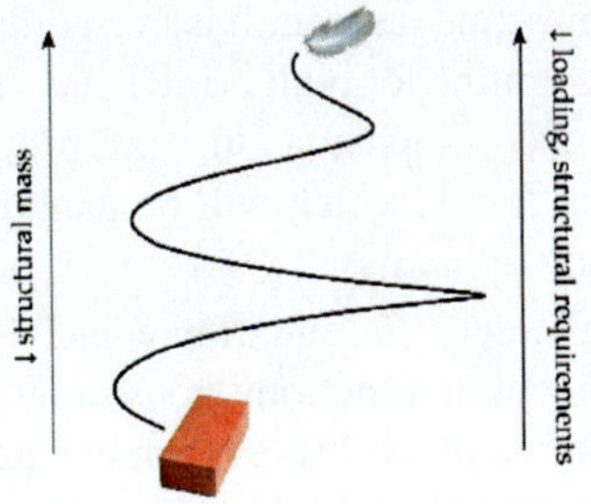

Fig. 1: Virtuous Circle of Lightweight Engineering Design

Efficient structural design optimization is an effective tool in entering the Virtuous Circle of Lightweight Engineering Design. Gradient-based optimization using analytical sensitivity analysis is especially efficient. Although structural design optimization has been brought to maturity for linear elasto-static structures [1, 9, 17], questions remain in areas of multiphysics, dynamics and nonlinearity. Structural dynamics and multibody dynamics are briefly introduced and the analytical design sensitivities using direct differentiation derived. Structural dynamics covers structural analysis under dynamic loading and behavior. Mechanisms and mechanical systems consisting of multiple bodies connected by joints are modeled and analyzed with multibody dynamics.

2 Sensitivity analysis of nonlinear dynamic systems

The centerpiece of efficient gradient-based design optimization is the sensitivity analysis. The categorization of sensitivity analysis is shown in fig. 2. The semi-analytical approach combines aspects of numerical and analytical sensitivities, using the analytical derivative of the governing equation with numerical sensitivities of the system parameters, i.e. the system matrices, e.g. mass and stiffness matrices. In this work a semi-analytical approach will be introduced unifying the calculation between structural dynamics and rigid multibody dynamics and their sensitivities. In doing so, the following sections builds and expands on the work of [1, 16].

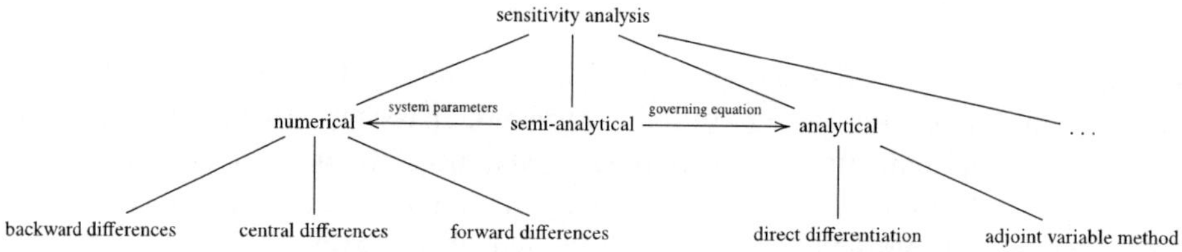

Fig. 2: Categorization for types of design sensitivity analysis

For transient analysis studied here, numerical time integration is needed with an integrated linear solver, which is called at every time step. In turn, sensitivity analysis must also be performed in this same fashion. It is most effective from an implementation view to use the same solving routine for both primal and sensitivity analysis. The unified code accepts matrices of higher dimensions to work properly. The flowchart and main building blocks are shown in fig. 3, which will be explained below for both structural dynamics and rigid multibody dynamics.

This will be shown using direct differentiation, which is best used (i.e. more efficient) when the number of optimization functions considered (i.e. sum of the number of objectives and number of constraints) is higher than the number of design variables. The adjoint variable method is more efficient when the number of design variables exceeds the number of optimization functions. This rule of thumb, though, does not consider the implementation complexity and effort.

3 Structural dynamics

3.1 Governing equation

In this section, the governing equation for structural dynamics will be introduced. In the second subsection, the sensitivity analysis will be derived and the nomenclature used throughout is introduced.

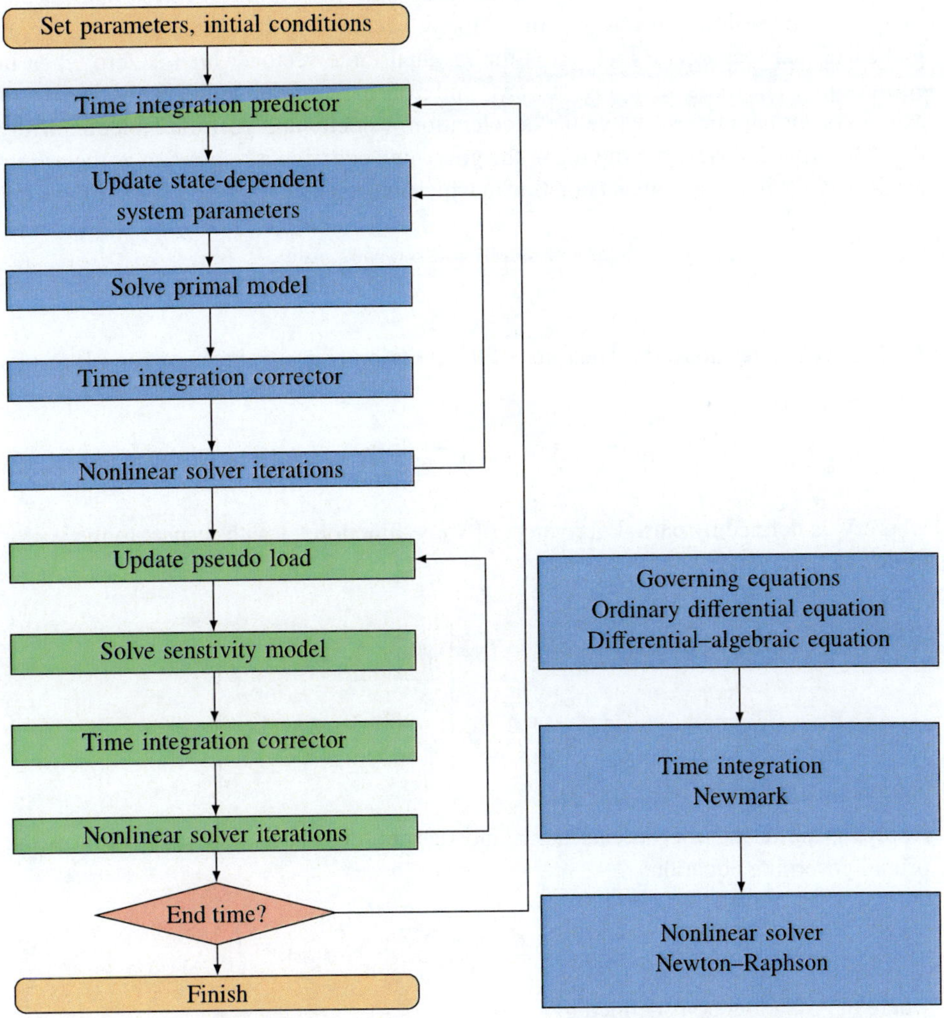

(a) Flowchart

(b) Generalized building blocks

Fig. 3: Solving procedure

3.1.1 Primary analysis The governing equation for structural dynamics is described here by the force equilibrium equation,

$$\underline{F}_R(t) = \underline{\underline{m}}\left(\underline{q}, t\right)\underline{\ddot{q}}(t) + \underline{\underline{d}}\left(\underline{q}, t\right)\underline{\dot{q}}(t) + \underline{\underline{k}}\left(\underline{q}, t\right)\underline{q}(t) - \underline{F}_{\text{ext}}\left(\underline{q}, t\right) = \underline{0}, \qquad (1)$$

where $\underline{\underline{m}}$ is the mass matrix, $\underline{\underline{d}}$ is the damping matrix, $\underline{\underline{k}}$ is the stiffness matrix, t is time, $\underline{q}$ is the position vector, $\underline{\dot{q}}$ is the velocity vector, $\underline{\ddot{q}}$ is the acceleration vector, $\underline{F}$ is the external load vector and $\underline{F}_R$ is the residual force vector, which is zero when in equilibrium. The mass, damping, stiffness and external force are collectively referred to as the system parameters, while the acceleration, velocity and position are collectively the state variables. Here the solving of the governing equation is referred to as the primal analysis. We will use a compact notation in which the dependence is not explicitly shown,

$$\underline{F}_R = \underline{\underline{m}}\,\underline{\ddot{q}} + \underline{\underline{d}}\,\underline{\dot{q}} + \underline{\underline{k}}\,\underline{q} - \underline{F}_{\text{ext}} = \underline{0}. \qquad (2)$$

3.1.2 Sensitivity analysis The direct differentiation of governing equation (2) results in

$$\underline{\underline{\nabla F}}_R = \underline{\underline{\underline{\nabla m}}}\,\underline{\ddot{q}} + \underline{\underline{m}}\,\underline{\underline{\nabla \ddot{q}}} + \underline{\underline{\underline{\nabla d}}}\,\underline{\dot{q}} + \underline{\underline{d}}\,\underline{\underline{\nabla \dot{q}}} + \underline{\underline{\underline{\nabla k}}}\,\underline{q} + \underline{\underline{k}}\,\underline{\underline{\nabla q}} - \underline{\underline{\nabla F}}_{\text{ext}} = \underline{\underline{0}}, \qquad (3)$$

where ∇ is defined by partial derivative of some function $(\cdot)$ with respect to the vector of design variables $\underline{x}$,

$$\underline{\underline{\nabla(\cdot)}} = \nabla_{\underline{x}}\underline{(\cdot)} = \frac{\mathrm{d}\underline{(\cdot)}}{\mathrm{d}\underline{x}} = \begin{bmatrix} \frac{\mathrm{d}(\cdot)_1}{\mathrm{d}x_1} & \frac{\mathrm{d}(\cdot)_1}{\mathrm{d}x_2} & \cdots & \frac{\mathrm{d}(\cdot)_1}{\mathrm{d}x_m} \\ \frac{\mathrm{d}(\cdot)_2}{\mathrm{d}x_1} & \frac{\mathrm{d}(\cdot)_n}{\mathrm{d}x_2} & \cdots & \frac{\mathrm{d}(\cdot)_2}{\mathrm{d}x_m} \\ \vdots & \vdots & \ddots & \vdots \\ \frac{\mathrm{d}(\cdot)_n}{\mathrm{d}x_1} & \frac{\mathrm{d}(\cdot)_n}{\mathrm{d}x_2} & \cdots & \frac{\mathrm{d}(\cdot)_n}{\mathrm{d}x_m} \end{bmatrix}. \qquad (4)$$

Applying the chain rule on each term and rearranging, leaves the same form as the primal governing equation,

$$\underline{\underline{\nabla F}}_R = \underline{\underline{m}}\,\underline{\underline{\nabla \ddot{q}}} + \underline{\underline{d}}\,\underline{\underline{\nabla \dot{q}}} + \underline{\underline{k}}\,\underline{\underline{\nabla q}} - \underline{\underline{F}}_{\text{pseudo}} = \underline{\underline{0}}, \qquad (5)$$

where the pseudo load is defined by

$$\underline{\underline{F}}_{\text{pseudo}} = \left(\underline{\underline{\partial F}}_{\text{ext}} + \underline{\underline{J}}_{F_{\text{ext}}}\underline{\underline{\partial q}}\right) - \left(\underline{\underline{\underline{\partial m}}} + \underline{\underline{J}}_m\,\underline{\underline{\partial q}}\right)\underline{\ddot{q}} - \left(\underline{\underline{\underline{\partial d}}} + \underline{\underline{J}}_d\,\underline{\underline{\partial q}}\right)\underline{\dot{q}} - \left(\underline{\underline{\underline{\partial k}}} + \underline{\underline{J}}_k\,\underline{\underline{\partial q}}\right)\underline{q}. \qquad (6)$$

The following nomenclature to show partials of the partial derivatives with respect to the design variables (to contrast them from total partial derivatives $\underline{\underline{\nabla(\cdot)}} = \frac{\mathrm{d}\underline{(\cdot)}}{\mathrm{d}\underline{x}}$):

$$\underline{\underline{\partial(\cdot)}} = \partial_{\underline{x}}\underline{(\cdot)} = \frac{\partial\underline{(\cdot)}}{\partial\underline{x}}. \qquad (7)$$

The Jacobians of the system variable are defined here by,

$$\underline{\underline{\mathrm{J}}}_{(\cdot)} = \mathrm{J}\left(\underline{(\cdot)}\right) = \frac{\partial \underline{(\cdot)}}{\partial \underline{q}}, \tag{8}$$

$$\underline{\underline{\dot{\mathrm{J}}}}_{(\cdot)} = \dot{\mathrm{J}}\left(\underline{(\cdot)}\right) = \frac{\partial \underline{(\cdot)}}{\partial \underline{\dot{q}}}, \tag{9}$$

$$\underline{\underline{\ddot{\mathrm{J}}}}_{(\cdot)} = \ddot{\mathrm{J}}\left(\underline{(\cdot)}\right) = \frac{\partial \underline{(\cdot)}}{\partial \underline{\ddot{q}}}. \tag{10}$$

For linear structural dynamics, the system parameters are independent of position and the total derivative is equal to the partial derivative, e.g. $\underline{\underline{\nabla m}} = \underline{\underline{\partial m}}$. The partials can be directly implemented for a fully analytical sensitivity analysis or calculated numerically, as the case is here, for a semi-analytical method.

3.2 Time integration

3.2.1 Primary analysis To solve this dynamic problem, the time integration will be shown here with Newmark-β method [14] with accelerations as the primary variable. This is carried out by solving eq. 2 for the accelerations at each time step (here i) using the predicted displacements and velocities,

$$^{i}\underline{q}_{\text{pred}} = {}^{i-1}\underline{q} + \Delta t\, {}^{i-1}\underline{\dot{q}} + \left(\frac{1}{2} - \beta\right) \Delta t^{2}\, {}^{i-1}\underline{\ddot{q}}, \tag{11}$$

$$^{i}\underline{\dot{q}}_{\text{pred}} = {}^{i-1}\underline{\dot{q}} + (1-\gamma)\, \Delta t\, {}^{i-1}\underline{\ddot{q}}, \tag{12}$$

where the left superscript i is the current time step, the left superscript $i-1$ is the previous time step, Δt is the time increment, while β and γ are the time integration constants. After updating the system parameters and calculating the acceleration, the effective mass and effective force are assembled,

$$\underline{\underline{m}}_{\text{eff}} = \underline{\underline{m}} + \gamma \Delta t \underline{\underline{d}} + \beta \Delta t^{2} \underline{\underline{k}}, \tag{13}$$

$$\underline{F}_{\text{eff}} = \underline{F}_{\text{ext}} - \underline{\underline{d}}\, {}^{i}\underline{\dot{q}}_{\text{pred}} - \underline{\underline{k}}\, {}^{i}\underline{q}_{\text{pred}}. \tag{14}$$

and then the following equation is solved for acceleration,

$$\underline{\underline{m}}_{\text{eff}}\, {}^{i}\underline{\ddot{q}} = \underline{F}_{\text{eff}}. \tag{15}$$

This acceleration is used to correct the predicted state variables,

$$^{i}\underline{q} = {}^{i}\underline{q}_{\text{pred}} + \beta \Delta t^{2}\, {}^{i}\underline{\ddot{q}}, \tag{16}$$

$$^{i}\underline{\dot{q}} = {}^{i}\underline{\dot{q}}_{\text{pred}} + \gamma \Delta t\, {}^{i}\underline{\ddot{q}}. \tag{17}$$

3.2.2 Sensitivity analysis The sensitivity analysis within the time integration is carried out completely analogously to the primary analysis, though using matrix-valued terms for the sensitivities instead of vectors of the system responses. For the prediction this becomes

$$^{i}\underline{\underline{\nabla q}}_{\text{pred}} = {}^{i-1}\underline{\underline{\nabla q}} + \Delta t\, {}^{i-1}\underline{\underline{\nabla \dot{q}}} + \left(\frac{1}{2} - \beta\right) \Delta t^{2}\, {}^{i-1}\underline{\underline{\nabla \ddot{q}}}, \tag{18}$$

$$^{i}\underline{\underline{\nabla \dot{q}}}_{\text{pred}} = {}^{i-1}\underline{\underline{\nabla \dot{q}}} + (1-\gamma)\, \Delta t\, {}^{i-1}\underline{\underline{\nabla \ddot{q}}}, \tag{19}$$

resulting in the following effective mass sensitivity and effective force sensitivity:

$$\underline{\underline{\underline{\nabla m}}}_{\text{eff}} = \underline{\underline{\underline{\nabla m}}} + \gamma \Delta t \underline{\underline{\underline{\nabla d}}} + \beta \Delta t^{2} \underline{\underline{\underline{\nabla k}}}, \tag{20}$$

$$\underline{\underline{\nabla F}}_{\text{eff}} = \underline{\underline{\nabla F}}_{\text{ext}} - \underline{\underline{\underline{\nabla d}}}\, {}^{i}\underline{\dot{q}}_{\text{pred}} - \underline{\underline{d}}\, {}^{i}\underline{\underline{\nabla \dot{q}}}_{\text{pred}} - \underline{\underline{\underline{\nabla k}}}\, {}^{i}\underline{q}_{\text{pred}} - \underline{\underline{k}}\, {}^{i}\underline{\underline{\nabla q}}_{\text{pred}}. \tag{21}$$

Then the following equation is solved for acceleration sensitivity:

$$\underline{\underline{\underline{\nabla m}}}_{\text{eff}}\, {}^{i}\underline{\underline{\nabla \ddot{q}}} = \underline{\underline{\nabla F}}_{\text{eff}}. \tag{22}$$

The predicted sensitivity values are then corrected, giving

$$^{i}\underline{\underline{\nabla q}} = {}^{i}\underline{\underline{\nabla q}}_{\text{pred}} + \beta \Delta t^{2}\, {}^{i}\underline{\underline{\nabla \ddot{q}}}, \tag{23}$$

$$^{i}\underline{\underline{\nabla \dot{q}}} = {}^{i}\underline{\underline{\nabla \dot{q}}}_{\text{pred}} + \gamma \Delta t\, {}^{i}\underline{\underline{\nabla \ddot{q}}}. \tag{24}$$

3.3 Nonlinear solver

3.3.1 Primary analysis In the general form, the system parameters (mass, damping, stiffness and force) are dependent of the state variables (acceleration, velocity and position), therefore a nonlinear solver is needed. The methodology here is general and this specific case will be shown with Newton–Raphson method,

$$\frac{\partial \underline{F}_{R}}{\partial \underline{\ddot{q}}} \Delta \underline{\ddot{q}} + \underline{F}_{R} = 0, \tag{25}$$

which requires then the Jacobian of the residual with respect to the accelerations, denoted here as $\underline{\underline{\ddot{\mathrm{J}}}}_{F_R}$.

3.3.2 Sensitivity analysis For the sensitivity analysis, the Newton–Raphson step is

$$\frac{\partial \underline{\underline{\nabla F}}_{R}}{\partial \underline{\underline{\nabla \ddot{q}}}} \Delta \underline{\underline{\nabla \ddot{q}}} = \underline{\underline{\nabla F}}_{R}, \tag{26}$$

where the Jacobian of the residual sensitivity with respect to the acceleration sensitivity is compactly denoted by $\underline{\underline{\underline{\underline{\ddot{\overset{\nabla}{\mathrm{J}}}}}}}_{\nabla F_R}$. Via expansion of the partial derivatives, it is found that this value is the Jacobian for the primary analysis times the four-dimensional identity matrix

$$\underline{\underline{\underline{\underline{\ddot{\overset{\nabla}{\mathrm{J}}}}}}}_{\nabla F_R} = \underline{\underline{\ddot{\mathrm{J}}}}_{F_R} \underline{\underline{\underline{\underline{e}}}}, \tag{27}$$

allowing the use with a conventional solver

$$\underline{\underline{\ddot{\mathrm{J}}}}_{F_R} \Delta \underline{\underline{\nabla \ddot{q}}} = -\underline{\underline{\nabla F}}_{R}. \tag{28}$$

4 Rigid multibody dynamics

Rigid multibody dynamics is the analysis method for problems of several bodies connected via kinematic constraints. Previous studies of sensitivity analysis with multibody analysis include [10, 2, 19]. Building upon these, the derivation is shown below categorized into the building blocks introduced above, the critical primal equations and their derivatives for the sensitivity analysis.

4.1 Governing equation

4.1.1 Primary analysis Analogously to sensitivity analysis with structural dynamics, we find the design sensitivities for multibody dynamics. The governing equations are shown here using the Lagrangian multiplier form leading to a system of differential–algebraic equations (DAE). In its most generic form, the index-3 DAE with holonomic constraints, is written as follows:

$$\underline{\underline{m}}\left(\underline{q},t\right)\,\underline{\ddot{q}}+\underline{\underline{\mathrm{J}}}_{\Phi}^{T}\,\underline{\lambda}=\underline{F}_{\text{ext}}\left(\underline{q},\underline{\dot{q}},t\right)+\underline{F}_{v}\left(\underline{q},\underline{\dot{q}},t\right), \tag{29}$$

$$\underline{\Phi}\left(\underline{q},t\right)=\underline{0}. \tag{30}$$

The multibody problem is implemented here using an index-1 differential algebraic equation to avoid numerical issues associated with the index-3 formulation. As such the constraint function is differentiated twice with respect to time, resulting in

$$\underline{\underline{m}}\,\underline{\ddot{q}}+\underline{\underline{\mathrm{J}}}_{\Phi}^{T}\,\underline{\lambda}=\underline{F}_{\text{ext}}+\underline{F}_{v}, \tag{31}$$

$$\underline{\underline{\mathrm{J}}}_{\Phi}\,\underline{\ddot{q}}=\underbrace{-\frac{\partial\underline{\dot{\Phi}}}{\partial t}-\underline{\underline{\mathrm{J}}}_{\dot{\Phi}}\,\underline{\dot{q}}}_{\underline{F}_{c}}, \tag{32}$$

where the velocity-based forces are put together. Rewriting this equation in matrix form gives

$$\underline{F}_{R}=\begin{bmatrix}\underline{\underline{m}} & \underline{\underline{\mathrm{J}}}_{\Phi}^{T}\\ \underline{\underline{\mathrm{J}}}_{\Phi} & \underline{\underline{0}}\end{bmatrix}\begin{bmatrix}\underline{\ddot{q}}\\ \underline{\lambda}\end{bmatrix}-\begin{bmatrix}\underline{F}_{\text{ext}}+\underline{F}_{v}\\ \underline{F}_{c}\end{bmatrix}=\underline{0}. \tag{33}$$

4.1.2 Sensitivity analysis The sensitivity analysis of the governing equation (33) is again carried out with direct differentiation, resulting in

$$\underline{\underline{\nabla F}}_{R}=\begin{bmatrix}\underline{\underline{m}} & \underline{\underline{\mathrm{J}}}_{\Phi}^{T}\\ \underline{\underline{\mathrm{J}}}_{\Phi} & \underline{\underline{0}}\end{bmatrix}\begin{bmatrix}\underline{\underline{\nabla\ddot{q}}}\\ \underline{\underline{\nabla\lambda}}\end{bmatrix}-\underline{\underline{F}}_{\text{pseudo}}=\underline{\underline{0}}, \tag{34}$$

where the pseudo load is a function of the total derivatives of the system parameters. It is important to note that this, as with structural dynamics, has the same form as the primal problem, enabling the use of the same solving procedure.

4.2 Time integration

The time integration and its sensitivity follows the pattern introduced above for structural dynamics. For rigid multibody dynamics, the updates are carried out from the accelerations $\ddot{\underline{q}}$ and is unnecessary for the Lagrangian multipliers from the kinematic constraint equations $\underline{\lambda}$. For the sensitivity analysis, the acceleration sensitivities $\underline{\underline{\nabla \ddot{q}}}$ are used to update velocity and position sensitivities, as with structural dynamics, see § 3.2.

4.3 Nonlinear solver

4.3.1 Primary analysis The nonlinear solver is applied to the sensitivity analysis in multibody dynamics analogously to structural dynamics. The primary variables, i.e. those for which we are solving for, are acceleration and the Lagrangian multiplier of the kinematic constraints $\left[\, \ddot{\underline{q}}\ \underline{\lambda} \,\right]^T$, giving

$$\left[\overset{\ddot{}}{\underline{\underline{\mathrm{J}}}}_{F_R}\ \overset{\lambda}{\underline{\underline{\mathrm{J}}}}_{F_R}\right]\begin{bmatrix}\Delta\ddot{\underline{q}}\\ \Delta\underline{\lambda}\end{bmatrix} + \underline{F}_R = \underline{0}, \tag{35}$$

where

$$\overset{\lambda}{\underline{\underline{\mathrm{J}}}}_{F_R} = \frac{\partial \underline{F}_R}{\partial \underline{\lambda}}. \tag{36}$$

4.3.2 Sensitivity analysis For the sensitivity analysis, the Newton–Raphson step is

$$\left[\overset{\ddot{\nabla}}{\underline{\underline{\underline{\mathrm{J}}}}}_{\nabla F_R}\ \overset{\nabla\lambda}{\underline{\underline{\underline{\mathrm{J}}}}}_{\nabla F_R}\right]\begin{bmatrix}\Delta\underline{\underline{\nabla\ddot{q}}}\\ \Delta\underline{\underline{\nabla\lambda}}\end{bmatrix} + \underline{\underline{\nabla F}}_R = \underline{0}, \tag{37}$$

where

$$\overset{\nabla\lambda}{\underline{\underline{\underline{\mathrm{J}}}}}_{\nabla F_R} = \frac{\partial \underline{\underline{\nabla F}}_R}{\partial \underline{\underline{\nabla\lambda}}}. \tag{38}$$

As with structural dynamics, the expansion of the partial derivatives results in the Jacobian for the primary analysis times the four-dimensional identity matrix,

$$\overset{\ddot{\nabla}}{\underline{\underline{\underline{\mathrm{J}}}}}_{\nabla F_R} = \overset{\ddot{}}{\underline{\underline{\mathrm{J}}}}_{F_R}\underline{\underline{\underline{\underline{e}}}}, \tag{39}$$

$$\overset{\nabla\lambda}{\underline{\underline{\underline{\mathrm{J}}}}}_{\nabla F_R} = \overset{\lambda}{\underline{\underline{\mathrm{J}}}}_{F_R}\underline{\underline{\underline{\underline{e}}}}. \tag{40}$$

With this, the same solving routine can be used and the necessary Jacobians have already been calculated in the primal step.

5 Numerical example – Optimal design of a hydropower intake rack cleaning mechanism

Multibody dynamics is applied to the Tyrolean weir cleaning mechanism shown in fig. 4. A Tyrolean weir is a water intake system for hydroelectric power plants developed for rivers with steep slope and high sediment transport. Water flows over a weir into a canal, where it is collected and forwarded to the pipeline and the turbine. A trash rack prevents particles, e.g. stones and branches, from entering at the intake. The cleaning mechanism is installed to clear the rack of particles that block water from the intake. The mechanism consists of a hydraulically driven cleaning rack that forces particles from the trash rack and these are then washed away by the water flow.

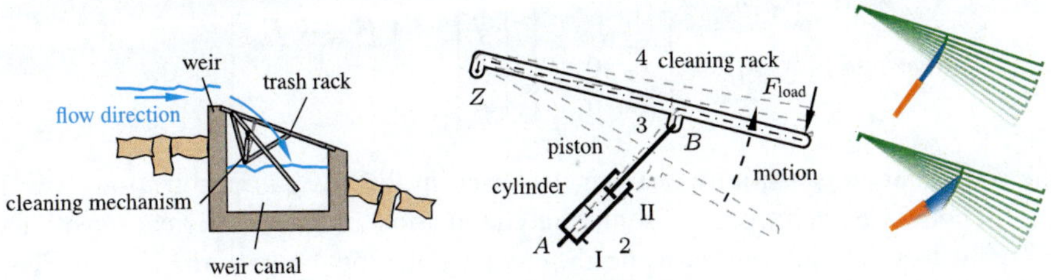

(a) Cross section of Tyrolean weir (b) Multibody system (c) Optimal designs

Fig. 4: Tyrolean weir cleaning mechanism

Design optimization of a Tyrolean weir cleaning mechanism is shown in [8] where the mechanism is modeled as planar multibody system and the sensitivities are computed with a reduced time integration method. Here, the mechanism is modeled in three-dimensional space and the sensitivity analysis is performed with the direct differentiation method shown in § 4. The Newmark-β method is used for the time integration and Newton–Raphson iterations are performed for both the primal and sensitivity analyses.

The mechanism consists of three moving bodies: the cylinder, the piston rod and the cleaning rack. These bodies are connected by three revolute joints and one prismatic joint. For the design optimization, three geometric and one hydraulic design variables are used to define the mechanism. The objective function is a multi-objective function with the minimization of the maximum force inside the joints that is approximated by the Kreiselmeier–Steinhauser function [11] and the pump flow rate multiplied by a weighting factor. Lower forces cause lower strains and stresses inside the components and enables to reduce the weight. Therefore, the energy consumption and the costs of the mechanism can be reduced. The optimization is limited by three constraint functions. The first constraint function defines an opening angle of at least 20 ° between the fully open cleaning rack and the fixed trash rack to allow the water to enter easily into the plant. The second constraint function defines a maximum cleaning time of 8 s in order to keep the energy losses caused by the cleaning process low. The third constraint function limits the stroke of the hydraulic cylinder to a maximum value of 500 mm, which was given by a limited budget for the hydraulic cylinder. The mathematical formulation of the optimization formulation is given by

$$
\begin{aligned}
&\min_{x \epsilon \chi} \mathcal{F}_{\mathrm{KS}}\left(\underline{F}_{\mathrm{res}}\right) + 0.01\, \dot{V}_S \\
&\text{such that} \quad \underline{g}\left(\underline{x}\right) = \begin{cases} \frac{-20\,^{\circ}}{\theta^{(4)}(t_0)} - 1 \\ \frac{r_{hc}(t_F)}{500\,\mathrm{mm}} - 1 \\ \frac{t_F}{8\,\mathrm{s}} - 1 \end{cases} \\
&\text{where} \quad \underline{x} = \left[\theta_{\overline{\mathrm{AZ}}}\, \ell_{\overline{\mathrm{AZ}}}\, \ell_{\overline{\mathrm{BZ}}}\, \dot{V}_S\right]^T \\
&\text{governed by} \quad \underline{F}_{\mathrm{R}} = \begin{bmatrix} \underline{\underline{m}} & \underline{\underline{J}}_{\Phi}^T & \underline{\underline{0}} \\ \underline{\underline{J}}_{\Phi} & \underline{\underline{0}} & \underline{\underline{0}} \\ \underline{\underline{0}} & \underline{\underline{0}} & \underline{\underline{e}} \end{bmatrix} \begin{bmatrix} \ddot{\underline{q}} \\ \underline{\lambda} \\ \dot{\underline{p}} \end{bmatrix} - \begin{bmatrix} \underline{F}_{\mathrm{ext}} + \underline{F}_v \\ \underline{F}_c \\ \underline{h} \end{bmatrix} = \underline{0}.
\end{aligned}
\tag{41}
$$

The semi-analytic sensitivity analysis was used in the design optimization. Tab. 1 and fig. (5) compare the introduced semi-analytic method and the numerical sensitivity method with forward differences using different values for the perturbation Δx. The sensitivities of the objective as well as of constraints 1 and 2 show convergence of the numerical sensitivities to the semi-analytical value with ever smaller perturbations. Conversely, this is not the case with the sensitivity of the third constraint (time of operation). Therefore, no perturbation value gives satisfactory results. Further, the direct sensitivity method drastically reduced the computation effort from $n_x + 1$ system evaluations (ca. 25 min per evaluation) to two system evaluations. Thus the theoretical speedup was achieved.

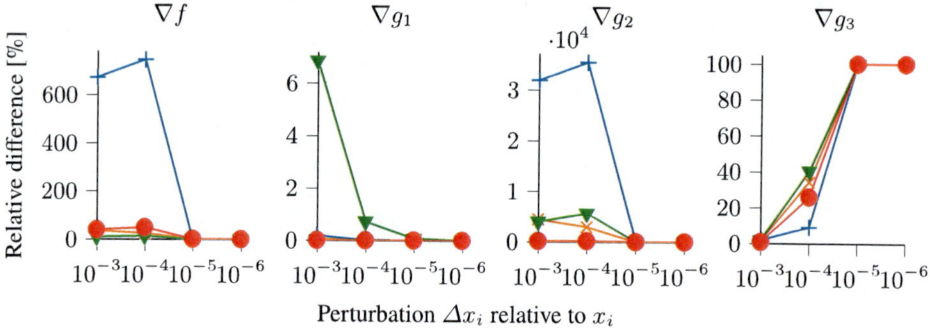

Fig. 5: Relative difference of numerical sensitivities with respect to semi-analytical sensitivities

Multiple optimization runs with the second-order algorithm NLPQLP [4, 15] from different start designs were performed. The optimizations converged to two different optima, which turned out to be equal from a qualitative point of view. The found optima are shown in fig. 4c. They can be considered as a mirrored design were the values of the lengths $\ell_{\overline{\mathrm{AZ}}}$ and $\ell_{\overline{\mathrm{BZ}}}$ are interchanged, leading to the same maximum joint force without violating the constraint functions.

Table 1: Comparison of semi-analytical and numerical sensitivity analysis calculated at $\underline{x} = [-1.22173,\ 1698.25,\ 1450.59,\ 493913]^T$

	semi-analytical	numerical			
Δx	–	$\underline{x} \cdot 10^{-3}$	$\underline{x} \cdot 10^{-4}$	$\underline{x} \cdot 10^{-5}$	$\underline{x} \cdot 10^{-6}$
t_{calc} [hh:mm:ss][1]	00 : 50 : 01	02 : 04 : 10	02 : 05 : 14	02 : 04 : 19	02 : 07 : 53
$\underline{\nabla f}$	$\begin{bmatrix} 1259.48 \\ -9.68942 \\ -19.7972 \\ 0.0161638 \end{bmatrix}$	$\begin{bmatrix} -7204.76 \\ -6.28831 \\ -17.9277 \\ 0.00995745 \end{bmatrix}$	$\begin{bmatrix} -8135.57 \\ -7.42888 \\ -17.161 \\ 0.00841248 \end{bmatrix}$	$\begin{bmatrix} 1267.79 \\ -9.68495 \\ -19.8033 \\ 0.0161657 \end{bmatrix}$	$\begin{bmatrix} 1265.29 \\ -9.68196 \\ -19.801 \\ 0.0161629 \end{bmatrix}$
$\underline{\nabla g_1}$	$\begin{bmatrix} 1.50778 \\ -0.000539986 \\ -1.255 \cdot 10^{-5} \\ 0.0 \end{bmatrix}$	$\begin{bmatrix} 1.50501 \\ -0.000540256 \\ -1.341 \cdot 10^{-5} \\ 0.0 \end{bmatrix}$	$\begin{bmatrix} 1.50751 \\ -0.000540013 \\ -1.264 \cdot 10^{-5} \\ 0.0 \end{bmatrix}$	$\begin{bmatrix} 1.50776 \\ -0.000539989 \\ -1.256 \cdot 10^{-5} \\ 0.0 \end{bmatrix}$	$\begin{bmatrix} 1.50778 \\ -0.000539986 \\ -1.255 \cdot 10^{-05} \\ 0.0 \end{bmatrix}$
$\underline{\nabla g_2}$	$\begin{bmatrix} 0.00861134 \\ 2.509 \cdot 10^{-5} \\ -1.53 \cdot 10^{-5} \\ 2.003 \cdot 10^{-6} \end{bmatrix}$	$\begin{bmatrix} -2.74135 \\ 0.00112417 \\ 0.000585078 \\ -1.415 \cdot 10^{-8} \end{bmatrix}$	$\begin{bmatrix} -3.04703 \\ 0.000757833 \\ 0.000842538 \\ -5.163 \cdot 10^{-7} \end{bmatrix}$	$\begin{bmatrix} 0.00861104 \\ 2.509 \cdot 10^{-5} \\ -1.531 \cdot 10^{-5} \\ 2.003 \cdot 10^{-06} \end{bmatrix}$	$\begin{bmatrix} 0.00861358 \\ 2.509 \cdot 10^{-05} \\ -1.531 \cdot 10^{-05} \\ 2.003 \cdot 10^{-06} \end{bmatrix}$
$\underline{\nabla g_3}$	$\begin{bmatrix} -2.81437 \\ 0.00111765 \\ 0.000613756 \\ -2.012 \cdot 10^{-6} \end{bmatrix}$	$\begin{bmatrix} -2.76248 \\ 0.00110408 \\ 0.000603204 \\ -2.025 \cdot 10^{-6} \end{bmatrix}$	$\begin{bmatrix} -3.06942 \\ 0.00073605 \\ 0.000861719 \\ -2.531 \cdot 10^{-6} \end{bmatrix}$	$\begin{bmatrix} 0.0 \\ 0.0 \\ 0.0 \\ 0.0 \end{bmatrix}$	$\begin{bmatrix} 0.0 \\ 0.0 \\ 0.0 \\ 0.0 \end{bmatrix}$

[1] PC with Intel® Core™ i7-8700 CPU @ 3.2 GHz × 12, 31.3 GiB

6 Conclusion

Cast within the Virtuous Circle of Lightweight Engineering Design, a unified approach, including a simplified nomenclature, was introduced for the sensitivity analysis of structural dynamics and rigid multibody dynamics. The analysis has been divided in three elementary building blocks and the respective equations are derived and implemented. These are then integrated into a gradient-based design optimization framework, and the optimization results are shown. The advantages in speedup and accuracy are all shown.

Although successful in the example shown, this method has the limitations common with gradient-based optimization. Nonsmooth, discontinuous and bifurcated optimization functions may cause convergence problems, especially when in the objective function. This can partially be avoided by using the lightweight engineering design formulation where the mass is the objective function, which a smooth and continuous function. Ill-conditioned optimization problems can be successfully handled with non-gradient algorithms (see e.g. [6, 5, 12, 13]) and approximation methods (see e.g. [3, 18, 7]).

The logical extension of this work is to derive and implement the equations for the adjoint variable method. While direct differentiation has advantages in its implementation and generality, the adjoint variable method has the edge of computational effort when the number of design variables is larger than the number of optimization functions (number of objectives plus the number of constraints). This becomes more apparent with great numbers of design variables as the case with shape and topology optimization.

Acknowledgments

This work is supported by the project CRC 2017 TN2091 doloMULTI *Design of Lightweight Optimized structures and systems under MULTIdisciplinary considerations through integration of MULTIbody dynamics in a MULTIphysics framework* funded by the Free University of Bozen-Bolzano.

References

1. Baier, H., Seeßelberg, C., Specht, B.: Optimierung in der Strukturmechanik. Vieweg (1994). https://doi.org/10.1007/978-3-322-90700-4
2. Bestle, D.: Analyse und Optimierung von Mehrkörpersystemen. Springer (1994). https://doi.org/10.1007/978-3-642-52352-6
3. Boursier Niutta, C., Wehrle, E.J., Duddeck, F., Belingardi, G.: Surrogate modeling in design optimization of structures with discontinuous responses: A new approach for ill-posed problems in crashworthiness design. Structural and Multidisciplinary Optimization **57**(5), 1857–1869 (mar 2018). https://doi.org/10.1007/s00158-018-1958-7
4. Dai, Y.H., Schittkowski, K.: A sequential quadratic programming algorithm with non-monotone line search. Pacific Journal of Optimization **4**, 335–351 (2008)
5. Duddeck, F.: Multidisciplinary optimization of car bodies. Structural and Multidisciplinary Optimization **35**(4), 375–389 (2008). https://doi.org/10.1007/s00158-007-0130-6

6. Duddeck, F., Hunkeler, S., Lozano, P., Wehrle, E., Zeng, D.: Topology optimization for crashworthiness of thin-walled structures under axial impact using hybrid cellular automata. Structural and Multidisciplinary Optimization **54**(3), 415–428 (2016). https://doi.org/10.1007/s00158-016-1445-y
7. Forrester, A., Sóbester, A., Keane, A.: Engineering design via surrogate modelling: A practical guide. In: AIAA Progress in Astronautics and Aeronautics (2008)
8. Gufler, V., Wehrle, E., Vidoni, R.: Multiphysical design optimization of multibody systems: Application to a Tyrolean weir cleaning mechanism. In: Advances in Italian Mechanism Science. pp. 459–467. Springer (2021). https://doi.org/10.1007/978-3-030-55807-952
9. Haftka, R.T., Gürdal, Z.: Elements of structural optimization. Kluwer, 3 edn. (1992). https://doi.org/10.1007/978-94-011-2550-5
10. Haug, E.J., Arora, J.S.: Applied optimal design: Mechanical and structural systems. John Wiley & Sons (1979)
11. Kreisselmeier, G., Steinhauser, R.: Systematic control design by optimizing a vector performance index. IFAC Proceedings Volumes **12**, 113–117 (1979). https://doi.org/10.1016/b978-0-08-024488-4.50022-x
12. Kurtaran, H., Eskandarian, A., Marzougui, D., Bedewi, N.E.: Crashworthiness design optimization using successive response surface approximations. Computational Mechanics **29**(4–5), 409–421 (2002). https://doi.org/10.1007/s00466-002-0351-x
13. Langer, H.: Extended evolutionary algorithms for multiobjective and discrete design optimization of structure. Dr.-ing. diss., Lehrstuhl für Leichtbau, Technische Universität München (2005)
14. Newmark, N.M.: A method of computation for structural dynamics. Journal of the Eng. Mech. Div. **85**(3) (1959)
15. Schittkowski, K.: NLPQLP: A fortran implementation of a sequential quadratic programming algorithm with distributed and non-monotone line search. User's guide, version 3.1, Department of Computer Science, University of Bayreuth (2010)
16. Trier, S.D., Marthinsen, A., Sivertsen, O.I.: Design sensitivities by the adjoint variable method in nonlinear structural dynamics. In: 38th SIMS Simulation Conference, Trondheim (1996)
17. Vanderplaats, G.N.: Multidiscipline design optimization. Vanderplaats Research & Development (2007)
18. Xu, Q., Wehrle, E., Baier, H.: Surrogate-based modeling and optimization, chap. Knowledge-based surrogate modeling in engineering design optimization. Springer (2013)
19. Zhu, Y.: Sensitivity analysis and optimization of multibody systems. Ph.D. thesis, Virginia Polytechnic Institute and State University (2014)

Hard- and Software fusion for process monitoring during machining of fiber reinforced materials

Markus G. R. Sause[1][0000-0002-6477-0691], Florian F. Linscheid[1], Christian Oblinger[1], Sebastian O. Gade[2] and Sinan Kalafat[2]

[1] University of Augsburg, 86135 Augsburg, Germany
[2] BCMtec GmbH, 86159 Augsburg, Germany
markus.sause@mrm.uni-augsburg.de

Abstract.
Machining of fiber reinforced composites remains a challenging process at the end of the value chain. Malfunctions of machinery and bad surface quality cause a tremendous loss of value and need to be reduced to increase competitiveness in lightweight applications. Monitoring of machining steps can be carried out by many different techniques and strategies all with their unique benefits and drawbacks. These provide information about machining hours, wear status of the tool, potential malfunctions of the system and can estimate the quality of the machining process. This contribution presents an approach to fuse different sensing systems on the hard- and software side to combine the information of different systems that provide a consolidated basis for the analysis of the machine status, tool status and machining quality. To this end we present results from a sensor fusion approach to measure acoustic information during the machining and the software framework UHU that was developed to provide a blueprint for a real-time capable environment for CNC feedback control and machining quality documentation.

Keywords: CNC machining, carbon fiber reinforced polymers (CFRP), condition monitoring, data fusion.

1 Introduction

The current production landscape is characterized by various trends. On the one hand, companies are faced with the need to flexibly produce smaller batch sizes in order to meet the increasing demand for individualized products without loss of productivity. On the other hand, technical developments offer new, previously unknown and thus unused possibilities for process support and improvement [1].

Machining of fiber reinforced materials - essentially milling, turning, finishing and drilling - is a process step that is essential for the fabrication of the final product as none of the production technologies of fiber reinforced materials immediately results in the final shape or surface quality required. However, the requirements posed to machining of these materials are constantly increasing. Tool life should be increased, contours

S. Pfingstl et al. (eds.), *Proceedings of the Munich Symposium on Lightweight Design 2020*, https://doi.org/10.1007/978-3-662-63143-0_6

should be milled more precisely and faster, the load capacity of the tool should be increased and downtimes of machines should be minimized. This is especially the case in series production, where repetitive processes need to be continuously improved and economically optimized.

Hence there is a strong trend towards exploitation of process data acquired by sensing technologies during machining of fiber reinforced materials to address these needs. However, integration of new sensing systems is sometimes costly and existing data sources inside conventional CNC machines may sometimes prove already as efficient means to solve some problems. Nevertheless, there is a serious gain in results when combining different data sources from different sensing systems. We present some recent work out of the project "WiR Augsburg" to provide a sophisticated framework for fusion of sensing systems as well as data fusion steps to move from rudimentary data acquisition systems to decision making "smart" machinery exploiting state-of-the-art artificial intelligence algorithms.

2 Hardware fusion

As seen in figure 1, the typical data acquisition systems found for monitoring of machining operations can be categorized into roughly three main types. There is classical machine data from CNC machines or turn tables comprising the continuous recording of tool position, tool or work piece velocity, overall operation conditions and many more. Into the same category fall systems which are already used on a broad basis for the controlling of the machinery, such as force/torque control systems to measure cutting forces as well as to collect 3D or 6D information about the loads acting on the tool [2–5]. Typical for the hardware in this category is that these systems are readily integrated into the machine controller and results may be extracted directly via industrial bus systems such as Profinet or OPC/UA.

As second category one can identify the acoustical monitoring systems. In this context, this can methodologically be split into three different sub-categories of measurement systems that are already established. Namely, these are vibration analysis systems and monitoring systems for acoustic emission, whereas the latter may be subdivided further into burst and continuous acoustic emission. There are existent standalone solutions for each of those types which typically are a serious financial investment. They have shown much potential to provide additional insight in the machining process, yet may not easily be integrated in existing shop floor solutions when it comes to use of industrial bus systems. We recently proposed a hardware fusion solution [6, 7] to integrate the required sensing capabilities into one sensor system. This is attached to one data acquisition unit, whereas differentiation between signal interpretation for vibration analysis or acoustic emission monitoring is done on the software side as will be explained next.

As third category we can identify camera monitoring systems that move along with the tooling to inspect the machining operation. For most applications of machining of fiber reinforced materials, it was identified that these systems do not provide much use cases, as the mixture of abrasive dust particles and fibers as well as lubricants or water inhibit a good view during the process and may degrade the optics over time. Nevertheless, scenarios with post-processing inspection via optical systems may prove useful, given the optics are properly protected during machining and enough time is provided for fog clearance after machining [8, 9]. These systems are currently all considered custom-made equipment.

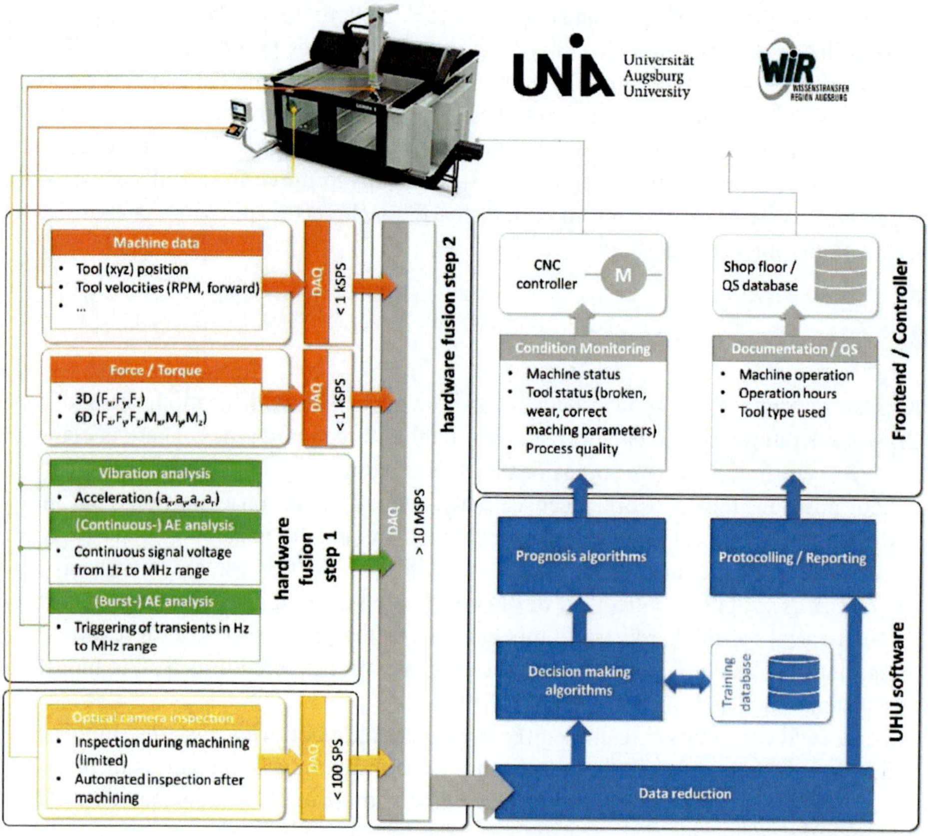

Fig. 1. Flow chart of hardware fusion steps for machining of fiber reinforced materials.

3 Software fusion

As an input from the different data acquisition systems one expects time series data packages, which are either one-dimensional (scalar values as function of time) or two-dimensional (vector or array values as function of time). In the present use case these are data streams of machine data, sensor voltage readings of acoustic sensors or image arrays of camera systems. Typically, these data streams are sampled at vastly different frequencies, so one key item is to provide a common time stamp for the whole data basis. Depending on the planned use of the data this determines the next processing steps. If the sole purpose of data acquisition is merely documentation, time lags do not cause much harm and real-time capability is not important as long as the system can dump the data during down-times of the machinery. However, as contrasting requirement, if online controlling of the CNC machine is intended (e.g. to change the tool when it is wearing out or to protect the work piece after tool breaks) real-time efficiency is crucial.

For the latter we developed the software framework Ultra-Highlevel-Ultrasonics (UHU) which is based on the aspect that the acoustic sensors provide the highest sampling rates considered (>10MSPS) and hence the software framework needs to be capable to process this data stream in real-time. Moreover, the UHU environment is intended to incorporate other data streams such as force/torque data streams or to integrate camera systems to operate on a common platform. The UHU approach sequentially processes packages of fixed time intervals of the data stream in a multi-core parallelized environment in a combination of edge and cloud computing as described in [10]. The core capabilities are based around a two-step feature extraction. For the first step, basic features in the time and frequency domain are extracted. A combination of those features for each package over a varying time span is then forwarded to decision-making algorithms (e.g. adaptive machine learning algorithms) which turn the low level features into high level predictions like tool wear status, location of defects, etc. Therefore, a backbone in form of a training database is linked via an interface. At the same time, other protocol and documentation steps may be delivered from the software or fed into a digital twin, so to complete the view of the current machining operation. For machining of fiber reinforced composites, the analysis of the acoustic signals also provides insight on the machining quality as well as on the tool wear status [2–5, 11–13], which can be delivered to a graphical user interface or can be directly reported into a shop floor QS system or database via an OPC/UA interface.

4 Conclusions

We proposed a software environment to combine data streams of different established monitoring methods for CNC-machining. The data intensity of the acoustic monitoring methods shows the highest demand regarding data throughput. Therefore, any hard- and software environment capable to deal with this data streams may in general be used to simultaneously acquire the other data streams resulting from the machine operation, from force/torque sensors or likewise instrumentation. The presented UHU environment was used to acquire such data during real production processes and will be continuously optimized for integration of advanced condition monitoring techniques and prognosis algorithms.

Acknowledgments. This paper presents results of the research project "WiR Augsburg – Wissenstransfer Region Augsburg", which is funded by the Federal Ministry of Education and Research (BMBF) within the program Innovative Hochschule.

References

1. Abele, E., Reinhart, G.: *Zukunft der Produktion - Herausforderungen, Forschungsfelder, Chancen - TUbiblio*. Carl Hanser Verlag, München (2011).
2. Scheer, C.: Überwachung des Zerspanungsprozesses mit geometrisch bestimmter Schneide durch Schallemissionsmessung, (2000).
3. Ferrari, G., Gómez, M.P.: *Correlation Between Acoustic Emission, Thrust and Tool Wear in Drilling*. Procedia Mater. Sci. **8**, 693–701 (2015).
4. Gómez, M.P., Hey, A.M., D'Attelis, C.E., Ruzzante, J.E.: *Assessment of Cutting Tool Condition by Acoustic Emission*. Procedia Mater. Sci. **1**, 321–328 (2012).
5. Gómez, M.P., Hey, A.M., Ruzzante, J.E., D'Attellisc, C.E.: *Tool wear evaluation in drilling by acoustic emission* In: Physics Procedia. pp. 819–825. Elsevier (2010).
6. Linscheid, F.F., Schwägerl, S., Sause, M.G.R.: *Evaluierung der Kombination von Schallemissionsanalyse und Prüfung mit geführten Wellen in einem gemeinsamen Sensornetzwerk* In: 22. Kolloquium Schallemission und 3. Anwenderseminar Zustandsüberwachung mit geführten Wellen. , Karlsruhe, Germany (2019).
7. Linscheid, F.F., Peter, T., Holzmann, C., Sause, M.G.R.: *Evaluierung eines Überwachungsszenarios durch Kombination von akustischen Zustandsüberwachungsmethoden in einem gemeinsamen Sensornetzwerk* In: DACH-Jahrestagung 2019: Zerstörungsfreie Materialprüfung. , Friedrichshafen, Germany (2019).
8. Xie, X.: *A Review of Recent Advances in Surface Defect Detection using Texture analysis Techniques*. ELCVIA Electron. Lett. Comput. Vis. Image Anal. **7**, 1 (2008).
9. Weimer, D., Scholz-Reiter, B., Shpitalni, M.: *Design of deep convolutional neural network architectures for automated feature extraction in industrial inspection*. CIRP Ann. **65**, 417–420 (2016).
10. Yang, C.-L., Sutrisno, H., Lo, N.-W., Chen, Z.-X., Wei, C.-C., Zhang, H.-W., Lin, C.-T., Wei, C.-L., Hsieh, S.-H.: *Streaming data analysis framework for cyber-physical system of metal machining processes* In: 2018 IEEE Industrial Cyber-Physical Systems (ICPS). pp. 546–551. IEEE.
11. Zhu, K.P., Wong, Y.S., Hong, G.S.: Wavelet analysis of sensor signals for tool condition monitoring: A review and some new results, (2009).
12. Marinescu, I., Axinte, D.: *A time-frequency acoustic emission-based monitoring technique to identify workpiece surface malfunctions in milling with multiple teeth cutting simultaneously*. Int. J. Mach. Tools Manuf. **49**, 53–65 (2009).
13. Krishnakumar, P., Rameshkumar, K., Ramachandran, K.I.: *Tool wear condition prediction using vibration signals in high speed machining (HSM) of Titanium (Ti-6Al-4V) alloy* In: Procedia Computer Science. pp. 270–275. Elsevier B.V. (2015).

Additive manufactured break-out cores for composite production: A case study with motorcycle parts

Manuel Biedermann[1], Max Widmer[1], Mirko Meboldt[1]

[1] Product Development Group Zurich pd|z, ETH Zürich, 8092 Zürich, Switzerland
manuel.biedermann@mavt.ethz.ch

Abstract. To produce hollow-shaped, lightweight composite structures made out of fiber reinforced polymers (FRP), many manufacturing processes require a shape-giving tooling in form of a core. Additive manufacturing (AM) offers the potential to fabricate such tools and production aids with increased geometric complexity and functionality at reduced costs and lead time. An AM core can remain inside the produced composite part and provide additional functionality such as the integration of metallic inserts. A core can also be removed from the final composite part to reduce the part mass. To enable the removal of a core, a promising approach is to use AM to design and produce a core in form of thin-walled shell that integrates breaking lines. After curing of the composite part, the breaking lines are used to break and disassemble the core into smaller patches, which are removed through an opening of the cured composite part. To stabilize the core shell during composite production, it is filled with a filler material such as salt. Although AM break-out cores offer many benefits, only a limited amount of works exists that study such cores. Therefore, this work contributes novel concepts for the design of AM break-out cores. The focus lies on the use of perforated and continuous breaking lines to enable a controlled fracture of cores. A case study demonstrates their application to produce parts of a motorcycle including the flow intake and tank structure. After the case study, the work discusses possible improvements and outlines future research directions.

Keywords: Composite Production, Tooling, Pull-able, Peel-able, Removable, Break-out, Core, Additive Manufacturing, Motorcycle, Flow Intake, Tank.

1 Introduction

To produce lightweight composite structures made out of fiber reinforced plastics (FRP), many manufacturing routes require tools such as cores, mandrels, and molds [1]. Processes for production of FRP parts include autoclave curing, filament winding, braiding, bladder molding or resin transfer molding. In case of autoclave curing, tools such as cores are used as shape-giving elements to drape and cure pre-impregnated fiber material (prepregs). During curing of the composite part, the core must withstand elevated pressure and temperature conditions and provide sufficient support and stability to ensure the accurate production of the composite part.

Tooling aids such as cores can either remain in the final composite part or be removed after the production process. Integrating a core inside a part can have advantages for part production, assembly, and operation. The integration of cores makes it possible

S. Pfingstl et al. (eds.), *Proceedings of the Munich Symposium on Lightweight Design 2020*, https://doi.org/10.1007/978-3-662-63143-0_7

to increase part stiffness, integrate metallic inserts and interfaces, and position production aids [2][3][4][5][6]. However, cores that remain inside the produced composite part can also have disadvantages. They may not fulfill any function during operation and increase the part mass as dead weight.

Therefore, a number of prior works investigate cores that are used for production but are removed from the final produced composite part [1]. To remove a core from a cured composite part, one approach is to fabricate the core out of special materials such as low-melting alloys or dissolvable materials. After autoclave curing, this allows to remove the core as a sacrificial tooling through a washout or melting process. However, such materials are usually quite expensive, and a complete removal may take a long time or require special equipment. Another approach is to use inflatable or foldable cores that form a shape-giving tool under pressure. Although the release of pressure enables a fast removal, the attainable tool shapes can be limited. A further approach is to use a multi-piece tooling, which is assembled from multiple modules. A drawback is that the attainable tool complexity may be limited or require an increased number of modules thereby increasing the effort for tool assembly and disassembly.

A number of studies investigate additive manufacturing process (AM) [7][8] such as binder jetting (BJ), fused deposition modeling (FDM), and selective laser sintering (SLS) to fabricate removable cores. Prior studies utilize BJ of water-soluble sand [9] or FDM of dissolvable materials [10]. AM processes are also used to produce removable break-out cores [11][12][13][14]. In this case, a core is designed as a thin-walled shell that integrates breaking lines. After curing of a composite part, breaking lines are used to break and disassemble the core into smaller patches. The patches can be removed through an opening of the part. Prior works fabricate polymer-based break-out cores using FDM and SLS. To stabilize the core during autoclave curing, the core is filled with salt as a cheap and temperature-resistant filler material with a low coefficient of thermal expansion [9][15].

The use of additive manufactured break-out cores with integrated breaking lines offers many advantages. Firstly, break-out cores made from SLS or FDM do not require special materials, equipment, or the need for a re-melting or dissolution process. Secondly, the use of AM makes it possible to cost-efficiently produce break-out cores for hollow composite parts with complex-shaped geometry at a reduced lead time. Thirdly, AM break-out cores can provide design features to insert functional elements such as load introductions that are co-cured with the composite part. This enables the production of highly functionally integrated and lightweight composite structures.

Despite these advantages only a limited amount of prior works investigates AM break-out cores with integrated breaking lines. Therefore, this work aims to further elaborate on the concept of AM break-out cores and their application. After providing background information on the basic manufacturing route in Sec. 2, the work presents novel ideas for the design of break-out cores in Sec. 3. The focus lies on the design of perforated and continuous breaking lines. Sec. 4 demonstrates the application for the production of parts of a motorcycle that are made from autoclave curing of FRP prepreg material. Sec. 5 discusses the results and outlines necessary improvements for the design and removal mechanism of AM break-out cores. Sec. 6 finishes the work with main conclusions and future research directions.

2 Additive manufactured break-out cores

This section describes the basic manufacturing route that is utilized to fabricate complex-shaped composite structures using additive manufactured break-out cores that serve as a sacrificial and removable tooling. The manufacturing route is based on a number of preliminary works [9][12][15]. The main steps are displayed in Fig. 1.

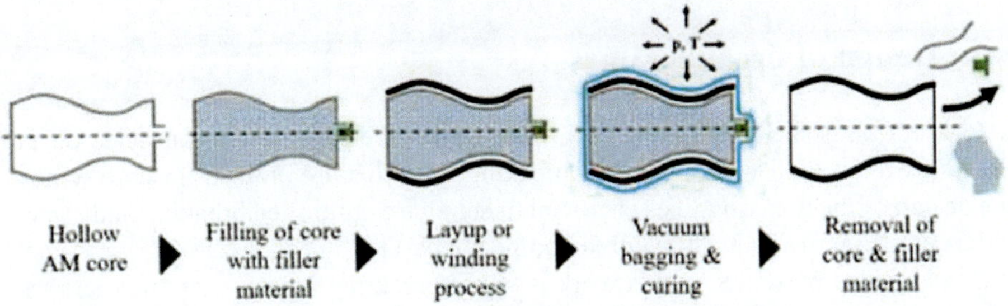

Fig. 1. Manufacturing route for the production of hollow composite structure using removable, AM break-out core that is filled with filler material during processing (adapted from [1]).

The first step consists in the design and fabrication of the core, which is used as a shape-giving tooling for the layup and curing of composite material. The core is designed as a hollow, thin-walled shell that integrates a number of breaking lines. Along the breaking lines, the core possesses a reduced wall thickness and notches to achieve a controlled fracture of the core along these lines. To fabricate the core, AM processes such FDM or SLS are employed. This allows producing a core with a complex-shaped geometry, which features breaking lines and can also integrate other design features, for example, to insert production aids or metallic load introductions and interfaces that are co-cured with the composite part.

In the next step the core is filled with a granular filler material that supports the shell during composite production and prevents a collapse or local deformation of the core. The filler material acts as a temporal support and is removed from the core after the curing steps. Prior works use salt as a cost-efficient filler material, which has a high temperature resistance, low coefficient of thermal expansion, and provides sufficient stiffness under compression. Based on its good flowability, salt also has the benefit to fill hollow spaces with intricate design features. After filling the core, it is enclosed with plugs. The filled core is further prepared by applying a release agent or release film. This prevents salt leaving the core and avoids a bonding between core and composite part, which allows separating and removing the core after the curing step.

In the following step composite material (e.g. prepreg) is draped on the core and cured in an autoclave. Alternatives for composite production are out-of-autoclave prepregs, filament winding, wet layup techniques, or resin transfer molding.

In case of autoclave curing, the next step includes bagging (e.g. adding of bleeder, breather, and release film) and curing under increased pressure and temperature to consolidate the prepreg. In the next step the part is unpacked, and the plugs of the core are opened to remove the filler material (e.g. washing out of salt).

To remove the core inside the composite part, the core is disassembled using the integrated breaking lines. The core is sequentially broken long the breaking lines into smaller patches, which are removed through openings of the cured composite part. For this purpose, the opening needs to be designed to provide sufficient access to crack and pull-out the patches. After core removal, the cured composite part is further post-processed. This includes milling of functional interfaces and finishing of part surfaces.

3 Design of breaking lines

AM processes such as FDM or SLS of polymers are suggested to fabricate the core as a thin-walled shell with integrated breaking lines that are utilized for core removal. The design of the breaking lines is crucial to achieve a controlled breaking and removal of the core from the inside of a cured composite part.

One key idea presented in this work is to use breaking lines in form of a perforated pattern of predetermined breaking points. As an example, Fig. 2 shows a tube-shaped, thin-walled core with perforated breaking lines. They partition the core into parallel-oriented patches. A close-up view displays main design parameters of the breaking lines. These include the gap width and bridge width of the perforated breaking line, wall thickness of the core, and reduced wall thickness along the breaking line.

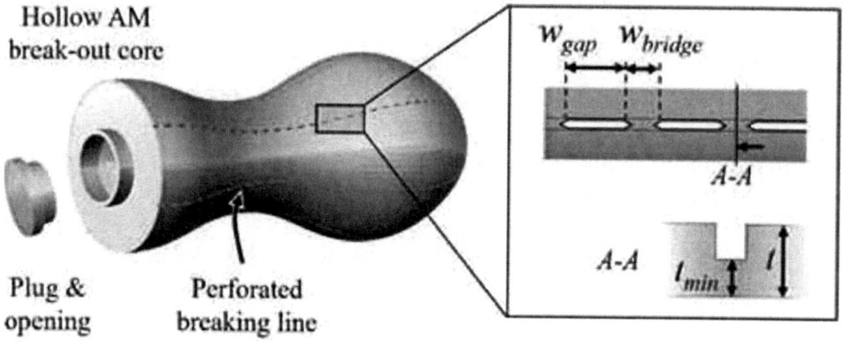

Fig. 2. Thin-walled break-out core with perforated breaking lines defined by of gap width w_{gap}, bridge width w_{bridge}, wall thickness t of core, and reduced wall thickness t_{min} at breaking line.

As a demonstration example, Fig. 3 (A) displays a simple test specimen. It is made out of ABS (acrylonitrile butadiene styrene) using FDM with a flat orientation on the build space. By pulling the patch a fracture is initiated, which propagates along the perforated breaking line as shown in Fig. 3 (B). The fracture mechanism corresponds to out-of-plane shearing and tearing (fracture mode III). The patch of the core is pulled using integrated leashes or peeled with pliers. The sequential pulling and removal of patches results in the disassembly of a core into smaller patches. These patches can be removed through an opening of a cured composite part. The opening of the composite part needs to provide sufficient access for core removal and pulling of patches.

Fig. 3. (A) Pulling of FDM-printed ABS patch with pliers; (B) Initiation and propagation of crack along predetermined, perforated breaking line.

The design freedom of FDM and SLS allows designing break-out cores with different shapes and configurations of breaking lines. Breaking lines can have various notch sizes, shapes (e.g. semicircular, U- and V-shaped notches), and perforation patterns. It is possible to use different layout strategies to partition the core into patches. Furthermore, one can leverage different fracture modes to pull and break the patches of a core. To decrease the number of patches that need to be removed, one promising approach is to design a core with a single continuous breaking line. As an example, Fig. 4 shows a test specimen, which integrates one continuous breaking line along a spiral-shaped pathway. Main design parameters are given in the figure and include the wall thickness of the core shell, wall thickness along the breaking line, and patch width.

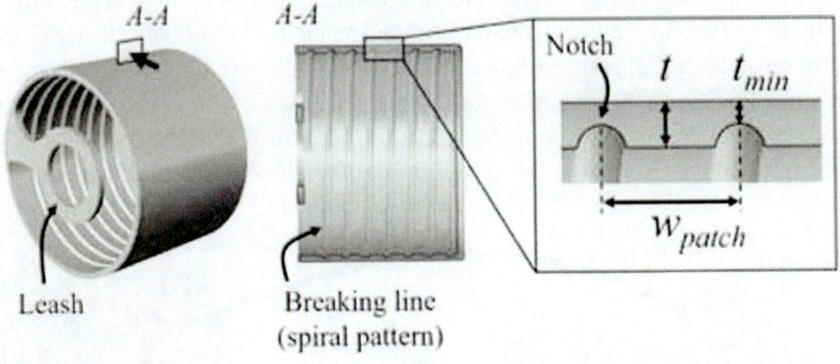

Fig. 4. Test specimen of break-out core with spiral-shaped breaking line defined by patch width w_{patch}, wall thickness t of core shell, and reduced wall thickness t_{min} along the breaking line.

Fig. 5 shows the test specimen with integrated spiral-shaped breaking line made out of polyamide using SLS. To break the core, a stress concentration and crack are initiated. By pulling the leash, the crack spreads along the breaking line. As shown in Fig. 5, the core is unwound as a continuous patch. The fracture mode along the U-shaped notch corresponds to opening (fracture mode I). With its reduced cross-sectional size, the patch shows flexible and spring-like properties. After the curing step, the idea is to use such a pulling mechanism of single patch can be used to remove the core from a cured composite part. The flexible properties of the patch make it possible to guide it through an opening of a cured composite part.

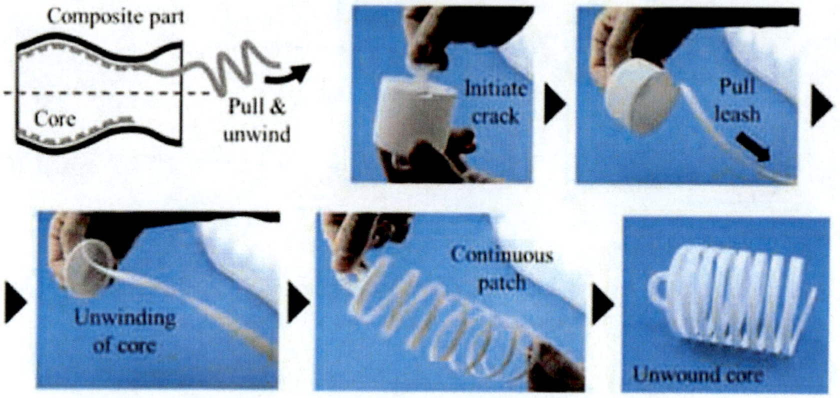

Fig. 5. Disassembly and breaking of SLS-made polyamide core by initiating crack and pulling leash to propagate crack along spiral-shaped breaking line and unwind core as a flexible patch.

4 Case study

The section demonstrates the use of AM break-out cores with perforated breaking lines to fabricate composite parts of a motorcycle, which is shown in Fig. 6. The examined parts include the flow intake and tank structure. The flow intake has the function to guide air into an airbox having an increased static pressure. The overpressure in the airbox increases the air flow rate into the engine and leads to an enhanced engine power. The tank provides a fuel reservoir of eight liters and is located in the rear frame of the motorcycle. To reduce the weight of the motorcycle, the flow intake and tank structure are produced using autoclave curing of carbon fiber reinforced polymers (CFRP) prepreg material. In the following, the production of the parts is described in detail.

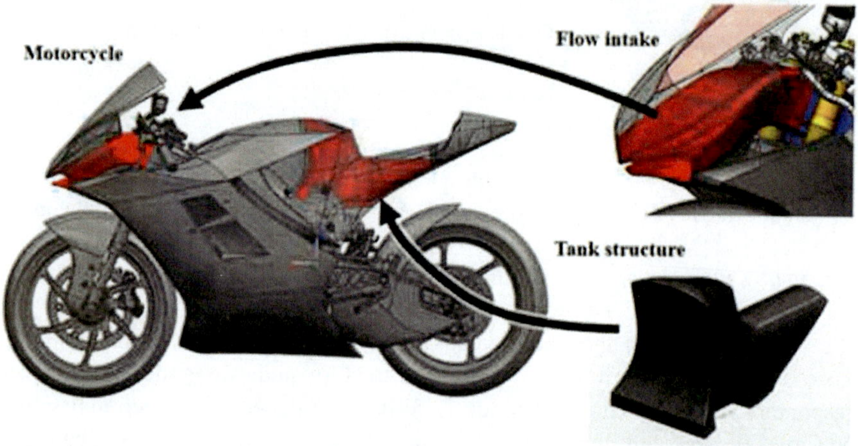

Fig. 6. Picture of motorcycle race bike developed by student team *ETH Moto Racing* (supervised by CMASLab of ETH Zurich) showing detailed view of flow intake and tank structure.

4.1 Flow intake

Fig. 7 depicts the production steps of the flow intake. Step 1 shows the thin-walled, break-out core, which has a wall thickness of $t = 2$ mm and contains a series of parallel oriented, perforated breaking lines ($t_{min} = 1$ mm, $w_{bridge} = 5$ mm, $w_{gap} = 7$ mm). FDM is used to fabricate the core out of ABS. The build orientation of the core is marked with an arrow. In step 2 prepreg sheets of twill fabric are cut for the layup, which consists of five layers on the top and bottom surfaces and seven on the side surfaces. The core is further prepared by applying a release film using glue. The release film prevents a bonding between the core and the prepreg and makes it possible to remove the core after curing the prepreg material. In step 3 the prepreg sheets are laid onto the core according to the layup plan. Further preparations include the application of additional release film, breather material, and vacuum bagging. To produce the flow intake, the core is not filled with filler material. Instead, given the two openings of the part the vacuum bag is set up in such way that it compresses the layup from two sides but does not deform and distort the shape of core. In step 4 curing of the prepreg is carried out at 84 °C for 12 h. After curing, the part is unpacked, and the core is fully removed by breaking the core into smaller patches using the breaking lines as shown in step 5. Some patches do not show a controlled fracture along the intended perforated breaking lines. In step 6 post-processing is applied such as milling and grinding of surfaces to get the finished composite part of the flow intake. The final flow intake part has a mass of 300 g.

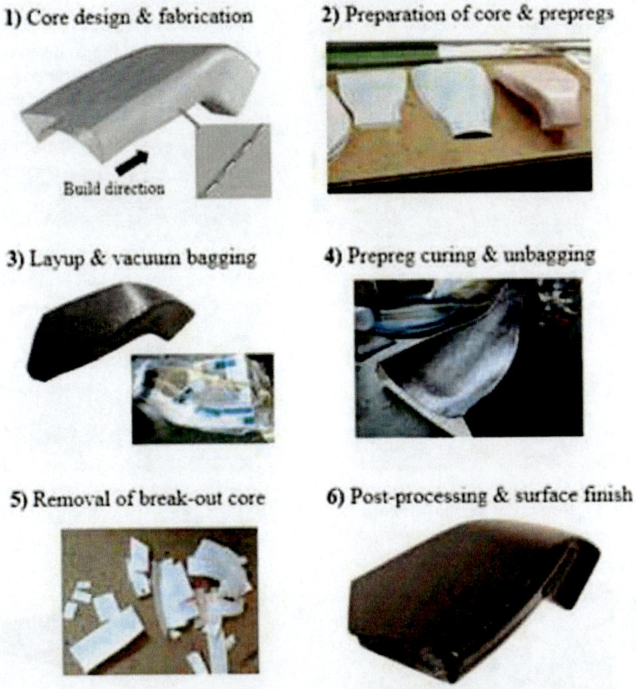

Fig. 7. Visualization of production steps of CFRP flow intake using a break-out core that is fabricated out of ABS using FDM, prepared with release film, used for layup and curing of prepreg material, and removed after composite processing through integrated breaking lines.

4.2 Tank structure

Fig. 8 shows the production steps of the tank. Step 1 consists in the design of the core with perforated lines. The core is designed with a wall thickness of $t = 2.4$ mm and several perforated breaking lines (t_{min} = 1.4 mm, w_{bridge} = 5 mm, w_{gap} = 7 mm) that subdivide the core into smaller patches. After curing of the composite part, the patches are removed through an opening that is located at the tank bottom. It provides access to the internal tank structure. In the final part, the opening is used to connect the tank with the fuel pump of the motorcycle. In step 2 the core is fabricated in ABS with FDM. The core is produced out of five sub-parts, which are printed in different build directions and joined through adhesive bonding. In step 3 release film is applied on the core and it is filled with salt. To fill the core with tightly packed salt, a vibrating table is used. Besides filling the thin-walled core with salt, the core design is locally enforced with stiffening ribs. These measures prevent the core from being indented during autoclave curing. After filling with salt, the opening at the core bottom is sealed with a plug.

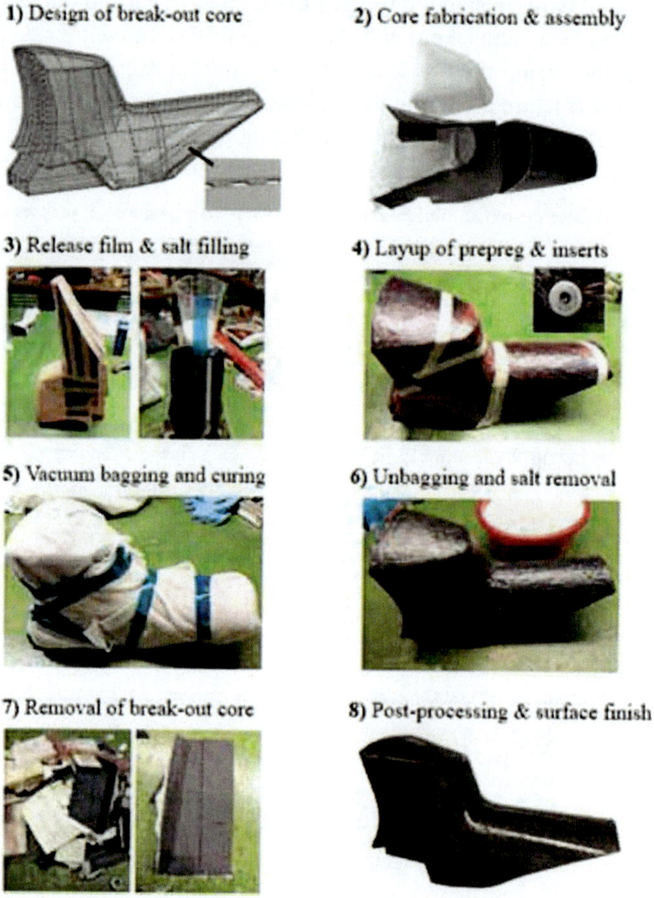

Fig. 8. Visualization of production steps of monolithic, hollow CFRP tank structure utilizing break-out core that is fabricated out of ABS using FDM, filled with salt as supporting material, used for layup and curing of prepreg material, and removed after processing of composite part.

As shown in step 4, metallic inserts are placed in the core. These are co-cured with the composite part and used to attach the tank to the frame. Also, the prepreg sheets are applied on the core. In step 5 release film and breather material are added before vacuum bagging. The composite part is cured at 84 °C for 12 h. It is then unpacked in step 6, and salt is removed by opening the core at the lower bottom of the tank and washing it out with water. Step 7 consists in breaking the core it into smaller patches using the integrated breaking lines and removing it through the opening. After core removal, the inner structure of the tank is cleaned with a solvent and sealed with chemically resistant epoxy. The final step 8 applies post-processing steps to the outer tank surfaces. Furthermore, functional surfaces are post-processed and milled (e.g. lower interface to fuel pump, upper opening towards fuel tank cap, integration of air vent). This leads to the final monolithic CFRP tank structure. The mass of the empty tank structure is 560 g.

5 Discussion

The case study demonstrates that AM break-out cores can be used for the production of complex-shaped, hollow composite parts. The salt-filled cores withstand elevated temperature and pressure conditions during autoclave curing. In both examined parts, the core is fully removed by breaking it along integrated breaking lines into smaller-sized patches. Main advantages of the approach are that it is accessible and does not require special materials, hardware or equipment. The approach is, however, still in an early development stage. At the current level of maturity, it is suitable for the production of prototypes, individualized composite structures, or small-series parts. Further improvements are required to achieve a controlled fracture and removal of cores.

For example, a number of patches do not show a fracture along the breaking lines. Instead a fracture occurs perpendicular to them and lies between printed layers defined by the build direction of FDM. Therefore, future works are needed to better understand the influence of AM (e.g. chosen AM process, parameters, material, build direction) on the fracture behavior. Processes such as SLS may be more suitable compared to FDM and reduce anisotropic effects caused by layer-wise manufacturing. To employ cores for increased curing temperatures, future studies may investigate the use of high temperature materials such as PEI (polyetherimide) or PEEK (polyether ether ketone).

Different concepts and mechanisms can be studied to achieve a robust and reliable removal and disassembly of AM cores. Future studies can investigate different designs and shapes for breaking lines and notches, other mechanisms for core removal (e.g. collapse of core under vacuum, use of wires for core pulling and fracture), multi-material AM (e.g. different materials at the breaking lines), or local modification of AM process parameters (e.g. porous material properties at breaking lines).

The design of cores with complex-shaped breaking line patterns (e.g. concept in Fig. 5) presents another bottleneck. Algorithms need to be developed that automate the design generation and layout of breaking lines based on a given input part geometry and constraints of a chosen AM process. In this respect, a simulation-based approach (e.g. extended finite element method, XFEM) may be used to predict and optimize a controlled facture and removal of AM cores.

6 Summary and outlook

This work examined additive manufactured break-out cores for the production of hollow-shaped, fiber reinforced composite structures. The cores are thin-walled and integrate breaking lines to enable their controlled fracture and removal from the inside of the cured composite part. To stabilize cores during autoclave curing, they are filled with a support material such as salt. This work contributed novel concepts for the design of breaking lines such as the use of perforated and continuous breaking lines. The work demonstrated the application for the flow intake and tank of a motorcycle. Future studies are necessary to achieve a fast and robust facture and disassemble of AM cores. To cost-efficiently implement the concept for series production, it is required to automate the design of cores, the removal and disassemble mechanism, and the layup of composite material (e.g. using filament winding, braiding, and fiber patch placement). Overall, a key idea presented in this work is to leverage the design freedom of AM processes to design structures in such a way that they exhibit a controlled fracture and breakage behavior. Besides AM break-out cores for composite structures, this concept may be applied for molds and tools for other forming and solidification processes such as casting, press forming, and injection molding as well as for mechanical parts that should rapture and collapse in a controlled behavior in response to a static or dynamic loading.

References

1. Türk, D.A.: Exploration and validation of integrated lightweight structures with additive manufacturing and fiber-reinforced polymers. PhD Thesis, ETH Zurich. (2017).
2. Türk, D.A., Kussmaul, R., Zogg, M., Klahn, C., Leutenecker-Twelsiek, B., Meboldt, M.: Composites Part Production with Additive Manufacturing Technologies. In: Procedia CIRP (2017). https://doi.org/10.1016/j.procir.2017.03.359.
3. Riss, F., Schilp, J., Reinhart, G.: Load-dependent optimization of honeycombs for sandwich components-new possibilities by using additive layer manufacturing. In: Physics Procedia. pp. 327–335. Elsevier B.V. (2014). https://doi.org/10.1016/j.phpro.2014.08.178.
4. Riß, F., Teufelhart, S., Reinhart, G.: Auslegung von Gitter- und Wabenstrukturen für die additive Fertigung. Light. Des. 6, 24–28 (2013). https://doi.org/10.1365/s35725-013-0187-7.
5. Kießling, R., Ihlemann, J., Pohl, M., Stommel, M., Dammann, C., Mahnken, R., Bobbert, M., Meschut, G., Hirsch, F., Kästner, M.: On the Design, Characterization and Simulation of Hybrid Metal-Composite Interfaces. Appl. Compos. Mater. 24, 251–269 (2017). https://doi.org/10.1007/s10443-016-9526-z.
6. Türk, D.A., Kussmaul, R., Zogg, M., Klahn, C., Spierings, A., Ermanni, P., Meboldt, M.: Additive manufacturing with composites for integrated aircraft structures. In: International SAMPE Technical Conference (2016).
7. Gibson, I., Rosen, D., Stucker, B.: Additive manufacturing technologies: 3D printing, rapid prototyping, and direct digital manufacturing, second edition. Springer New York (2015). https://doi.org/10.1007/978-1-4939-2113-3.
8. Türk, D.A., Brenni, F., Zogg, M., Meboldt, M.: Mechanical characterization of 3D printed polymers for fiber reinforced polymers processing. Mater. Des. 118, 256–265 (2017). https://doi.org/10.1016/j.matdes.2017.01.050.
9. Türk, D.A., Triebe, L., Meboldt, M.: Combining Additive Manufacturing with Advanced Composites for Highly Integrated Robotic Structures. In: Procedia CIRP. pp. 402–407. Elsevier B.V. (2016). https://doi.org/10.1016/j.procir.2016.04.202.

10.Türk, D.A., Einarsson, H., Lecomte, C., Meboldt, M.: Design and manufacturing of high-performance prostheses with additive manufacturing and fiber-reinforced polymers. Prod. Eng. 12, 203–213 (2018). https://doi.org/10.1007/s11740-018-0799-y.
11.Türk, D.A., Rüegg, F., Biedermann, M., Meboldt, M.: Design and manufacture of hybrid metal composite structures using functional tooling made by additive manufacturing. Des. Sci. 5, (2019). https://doi.org/10.1017/dsj.2019.16.
12.Kussmaul, R., Biedermann, M., Pappas, G.A., Jónasson, J.G., Winiger, P., Zogg, M., Türk, D.-A., Meboldt, M., Ermanni, P.: Individualized lightweight structures for biomedical applications using additive manufacturing and carbon fiber patched composites. Des. Sci. 5, (2019). https://doi.org/10.1017/dsj.2019.19.
13.Pallari, J.H.P.: WO2013050524A1 - Additive manufacturing of tiled objects, (2013).
14.Bremmer, J., Toni, D.M., Sauer, J.G., Lacko, R.A., Prieto, J.J., Foti, C.J.: WO2016153588A1 - Tools and processes for manufacturing parts employing additive manufacturnig, (2016).
15.Türk, D.A., Ebnother, A., Zogg, M., Meboldt, M.: Additive manufacturing of structural cores and washout tooling for autoclave curing of hybrid composite structures. J. Manuf. Sci. Eng. Trans. ASME. 140, (2018). https://doi.org/10.1115/1.4040428.

Neue Bauweisen von Wasserstoffdruckbehältern für die Integration in zukünftige Fahrzeugarchitekturen

Michael Ruf[1,2(✉)][0000-0002-6834-8121], Hans-Ulrich Stahl[1], Klaas Kunze[1], Swen Zaremba[2], Alexander Horoschenkoff[3], Thomas von Unwerth[4] und Klaus Drechsler[2]

[1] BMW AG, Abteilung für Wasserstoffversorgung, München, Deutschland
[2] Technische Universität München, Lehrstuhl für Carbon Composites, Garching, Deutschland
[3] Hochschule für angewandte Wissenschaften München, Verbundlabor Polymertechnik, München, Deutschland
[4] Technische Universität Chemnitz, Lehrstuhl für Alternative Fahrzeugantriebe, Chemnitz, Deutschland
Michael.G.Ruf@bmw.de

Kurzfassung. Auf dem globalen Automobilmarkt stellen Brennstoffzellenfahrzeuge im Vergleich zu Batterieelektrofahrzeugen immer noch ein Nischenprodukt dar. Neben einem dünnen Tankstellennetz, das sich erst im Aufbau befindet, ist ein hoher Fahrzeugpreis einer der Gründe dafür. Ein großes Potenzial für Kostenreduktion birgt die Nutzung gemeinsamer Fahrzeugarchitekturen für batterieelektrische sowie mit Wasserstoff betriebene Fahrzeuge, um Skaleneffekte in der Entwicklung und Produktion nutzen zu können. Die Nutzung gleicher Bauräume für die Antriebskomponenten und Energiespeicher ist dafür eine Grundvoraussetzung. Für zukünftige Brennstoffzellenfahrzeuge bedeutet dies eine Integration der Wasserstoffdruckbehälter in die flachen Batteriespeicherbauräume im Fahrzeugunterboden. Im Zuge dessen werden Tankkonzepte basierend auf zylindrischen Druckbehältern untersucht. Besondere Herausforderungen stellen dabei das Fertigungsverfahren sowie die Permeation dar. Eine erste Potenzialabschätzung für das Speichervolumen zeigt die Umsetzbarkeit kundenrelevanter Fahrzeugreichweiten. Als alternatives Konzept werden nahezu quaderförmige Druckbehälter mit Zugverstrebung im Inneren untersucht. Ein neuartiges Fertigungsverfahren wurde für die Herstellung entwickelt und anhand eines Baumusters validiert. Eine Potenzialabschätzung für die Speicherkapazität zeigt im Vergleich zur Variante mit zylindrischen Komponenten ein ca. 25 % höheres Wasserstoffvolumen. Die Technologie befindet sich allerdings noch im Forschungsstadium.

Stichwörter: Brennstoffzellenfahrzeug, Wasserstoff, Druckbehälter, Fahrzeugarchitektur, Tankintegration.

1 Motivation und Stand der Technik

1.1 Ökologischer und wirtschaftlicher Hintergrund der Untersuchungen

Im März 2019 beschloss das EU Parlament eine weitere Reduktion der CO2-Grenzwerte für die Neuwagenflotten der Automobilhersteller. Dabei soll der Ausstoß

S. Pfingstl et al. (eds.), *Proceedings of the Munich Symposium on Lightweight Design 2020*, https://doi.org/10.1007/978-3-662-63143-0_8

des Treibhausgases bis 2030 um weitere 37,5 % gegenüber den Werten, die es bis 2021 zu erreichen gilt, reduziert werden [1]. Um diese Ziele zu erfüllen, rücken für Automobilhersteller alternative Antriebe vermehrt in den Fokus. Eine Möglichkeit für den schadstofffreien Betrieb von Kraftfahrzeugen stellt ein elektrifizierter Antriebsstrang dar, dessen elektrische Energie durch einen Batteriespeicher oder eine mit Wasserstoff betriebene Brennstoffzelle bereitgestellt wird. Aus Kundensicht haben Brennstoffzellenfahrzeuge (engl. Fuel Cell Electric Vehicles, kurz FCEVs) gegenüber Batterieelektrofahrzeugen (engl. Battery Electric Vehicles, kurz BEVs) vor allem den Vorteil der kürzeren Betankungszeit (verglichen mit dem Ladevorgang), konnten sich aber aufgrund des aktuell dünnen Tankstellennetzes sowie der höheren Anschaffungskosten noch nicht durchsetzen [2].

1.2 Fahrzeugarchitekturen für zukünftige emissionsfreie Antriebsvarianten

Damit ein FCEV eine wettbewerbsfähige Alternative zum BEV wird, müssen vor allem die Herstellkosten gesenkt werden. Ein wesentliches Potenzial birgt die Steigerung der Stückzahl, durch welche die Entwicklungs- und Produktionskosten infolge von Skaleneffekten deutlich reduziert werden können [3]. Dieser Effekt kann auch bezogen auf die Fahrzeugarchitektur genutzt werden, indem in Elektrofahrzeugen unabhängig vom Energiespeicherkonzept gleiche Architekturen und Bauräume für die Antriebssystemkomponenten genutzt werden – ähnlich wie es heute in Diesel- und Benzinfahrzeugen umgesetzt ist. In aktuellen Brennstoffzellenfahrzeugen erfolgt die Speicherung des gasförmigen Wasserstoffs üblicherweise in einem, zwei oder drei zylindrischen carbonfaserverstärkten Druckbehältern mit vergleichsweise großen Durchmessern, welche im Mitteltunnel längs zur Fahrtrichtung und/oder vor und hinter der Fahrzeughinterachse quer zur Fahrtrichtung positioniert werden [3, 4]. Dagegen wird in Batterieelektrofahrzeugen der Hochvoltspeicher aufgrund seiner Größe, seines Gewichts sowie der Sicherheit im Crashfall typischerweise rein im Fahrzeugunterboden positioniert. Unter verschiedenen Integrationsstrategien besitzt dort der Sandwichboden das größte Potenzial hinsichtlich Modularisierung des Batteriespeichersystems sowie Ausnutzung von Baureihensynergien [2]. Die Dimensionen dieses quaderförmigen Bauraums erstrecken sich in der Länge zwischen den Fahrzeugachsen, in der Breite zwischen den Längsträgern und in der Höhe zwischen Fahrzeugunterbodenverkleidung und Fahrgastinnenraum (Fig. 1).

Fig. 1. Sandwichintegration des Hochvoltbatteriespeichers in den Fahrzeugunterboden eines BEVs beispielhaft dargestellt am *BMW i3* [5].

1.3 Geometrische Anpassung der Druckbehälterform als Lösung

Die aktuell quaderförmigen, flachen Batteriespeicherbauräume stellen für die Integration von Wasserstoffdruckbehältern eine besondere Herausforderung dar, da sich aus konstruktiver Sicht aufgrund des hohen Innendrucks vor allem kugel- oder zylinderförmige Geometrien für die Druckbehälter eignen, um eine möglichst gute Spannungsverteilung zu erreichen [3, 6]. Deshalb werden zwei neuartige Bauweisen für Speichersysteme von gasförmigem Wasserstoff untersucht, die derartige Bauraumdimensionen mit einer möglichst hohen volumetrischen und gravimetrischen Dichte füllen können.

Zum einen wird der Lösungsansatz verfolgt, die grundlegende, zylindrische Außengeometrie des Druckbehälters beizubehalten. Der quaderförmige Bauraum wird dabei mit mehreren Einzelelementen ausgefüllt, um durch die dreidimensionale Parkettierung einen gesteigerten Volumennutzungsgrad zu erreichen (Fig. 2, links). Dabei werden zwei Ausführungsvarianten unterschieden. Einerseits kann das Tanksystem differenziell aufgebaut werden, so dass die Zylinder für sich jeweils einen abgeschlossenen Druckbehälter darstellen und über ein daran angeschlossenes Leitungssystem miteinander verbunden werden. Andererseits können die zylindrischen Elemente beispielsweise durch eine integrale Fertigung direkt zu einem Druckbehälter verbunden werden, der durch eine mäanderförmige Anordnung der Einzelelemente den Bauraum füllt.

Im zweiten Lösungsansatz wird untersucht, die Geometrie des quaderförmigen Bauraums mit der Außenform des Druckbehälters nahezu abzubilden. Um aufgrund der hohen Druckbelastung im Inneren des Behälters eine Verformung zu vermeiden bzw. gleichmäßig und gering zu halten, werden die Ober- und Unterseite durch Zugstreben miteinander verbunden (Fig. 2, rechts). Die Anbindung der Zugstreben an die Außenwände des Druckbehälters spielt dabei eine wesentliche Rolle, da in diesem Bereich großenteils die Kräfte übertragen werden, die dem Behälterinnendruck entgegenwirken.

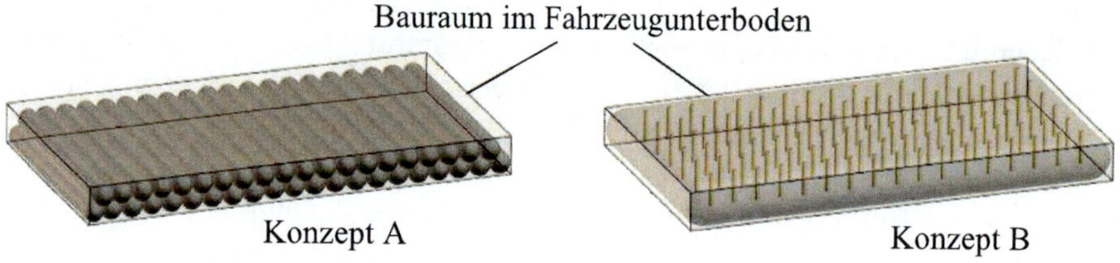

Fig. 2. Lösungsansätze zur Nutzung des Bauraums im Fahrzeugunterboden für die Speicherung des gasförmigen Wasserstoffs in zylindrischen Druckbehältern (links) sowie quaderförmigen Druckbehältern mit Zugverstrebungen (rechts).

1.4 Wasserstoffspeicher in aktuellen Brennstoffzellenfahrzeugen

Aktuell wird bei den Automobilherstellern in der Entwicklung von FCEVs auf bestehende Fahrzeugarchitekturen mit Verbrennungsmotorantrieb aufgesetzt, da eine eigene Baureihe aufgrund der noch relativ geringen Stückzahlen zu sehr hohen Entwicklungs- und Produktionskosten führen würde. Für die Integration der Antriebskomponenten ergeben sich im Brennstoffzellenfahrzeug demnach freie Bauräume vor und hinter der

Mittelachse sowie im Mitteltunnel durch den Wegfall des konventionellen Kraftstofftanks sowie des Getriebes und der Antriebswelle. Im FCEV können an diesen Positionen die Wasserstoffdruckbehälter integriert werden (Fig. 3). Die Bauräume erlauben in diesen Fällen verhältnismäßig große Durchmesser für die zylindrischen Kraftstoffspeicher. Das Brennstoffzellensystem wird in der Regel im Fahrzeugvorderwagen integriert, wo entsprechender Bauraum durch den Entfall des Verbrennungsmotors zur Verfügung steht.

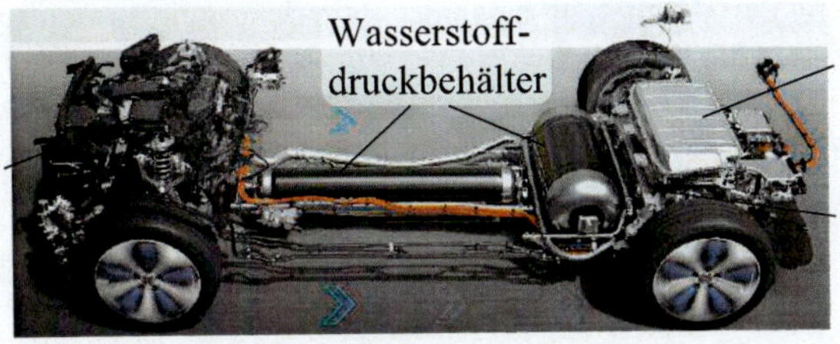

Fig. 3. Anordnung der Wasserstoffdruckbehälter in einem FCEV beispielhaft dargestellt am *Mercedes-Benz GLC F-Cell* [8].

Aktuelle Wasserstofftanks lassen sich den Typ IV Druckbehältern zuordnen, was bedeutet, dass der Behälter aus einem Kunststoffliner sowie einer Armierungsschicht aus carbonfaserverstärktem Kunststoff (CFK) besteht. Die Hauptfunktion des Liners ist dabei, den Tank gegenüber der Umwelt abzudichten und die Wasserstoffpermeation zu begrenzen. Über die Armierungsschicht wird die Lastaufnahme infolge der Innendruckbeanspruchung gewährleistet. Der Nennbetriebsdruck der Wasserstofftanks beträgt 700 bar.

2 Zylindrische Druckbehälter

2.1 Konzeptvarianten

Für die Integration von Wasserstoffspeichern in Architekturen von (Batterie-)Elektrofahrzeugen erweisen sich die Druckbehälterdimensionen aktueller FCEVs aufgrund ihres großen Durchmessers als nicht geeignet. Der Bauraum ermöglicht allerdings die Positionierung mehrerer zylindrischer Behälter nebeneinander in horizontaler Richtung oder bei ausreichend kleiner Dimensionierung zusätzlich übereinander. Durch die Verbindung untereinander werden die Zylinder zum Tanksystem des FCEVs zusammengeschlossen. In Patentschriften sowie Offenlegungsschriften zu deren Anmeldung sind Lösungsansätze für die Integration solcher Wasserstoffspeicher in den Fahrzeugunterboden bereits dargestellt [9-12]. Dabei lassen sich grundsätzlich zwei Arten der Tanksystemausführung und damit der Verbindung der einzelnen Zylinder unterscheiden. In einer differenziellen Ausführung stellt jeder Zylinder für sich einen eigenen Druckbehälter dar. Das Tanksystem besteht aus dem Zusammenschluss der einzelnen Behälter über ein Leitungssystem (Fig. 4, links), ähnlich wie es in aktuellen FCEVs ebenfalls der Fall ist. In einer integralen Ausführung sind die Zylinder so miteinander verbunden, dass sie als Einheit einen einzigen Druckbehälter bilden (Fig. 4, rechts).

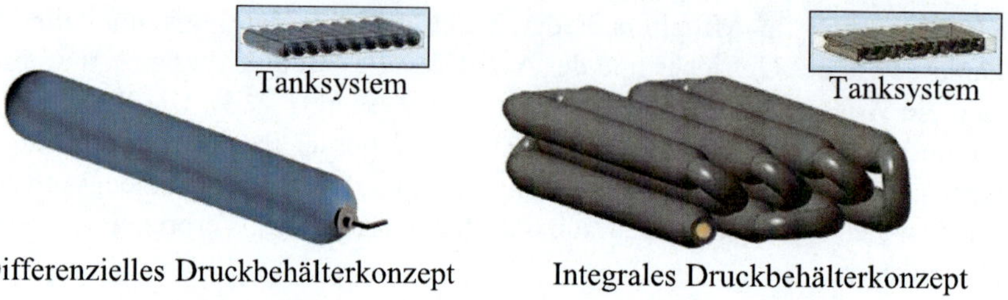

Fig. 4. Unterscheidung der Verbindung einzelner Zylinder zu einem Tanksystem in eine differenzielle sowie eine integrale Ausführung.

Zum heutigen Zeitpunkt gibt es für die dargestellten Konzepte kein Serienprodukt, das für die Anwendung im Brennstoffzellenfahrzeug geeignet und gleichzeitig zertifiziert ist. In einem durch das US-amerikanische Department of Energy geleiteten Forschungsprojekt wurden erste Grundlagenuntersuchungen zur integralen Bauweise durchgeführt. Anhand des Aufbaus erster Prototypen konnten das Fertigungskonzept sowie die mechanischen Eigenschaften validiert werden. Die zu Projektbeginn als Ziel festgelegte Leckagerate infolge der Wasserstoffpermeation durch die Druckbehälterwand wurde nicht erreicht und das Projekt aus diesem Grund nicht fortgesetzt [13].

2.2 Fertigungstechnologie und Permeation als zentrale Herausforderungen

Die Fertigungstechnologie spielt sowohl für heutige als auch zukünftige Druckbehälter eine wesentliche Rolle für deren Auslegung und konstruktive Gestaltung. Für aktuelle Wasserstofftanks hat sich das Wickelverfahren als Fertigungstechnologie zum Aufbringen der Armierungsschicht aus CFK etabliert [7]. Für eine beanspruchungsgerechte Gestaltung der Armierungsschicht werden die Fasern parallel in Umfangsrichtung und über Kreuz in Axialrichtung abgelegt (Fig. 5, links). Die fertigungstechnische Realisierung eines integralen Druckbehälterkonzepts ist im Wickelverfahren aufgrund der komplexen Bauteilgeometrie allerdings nur bedingt geeignet. Im Gegensatz dazu ermöglicht beispielsweise das Flechten eine integrale Fertigung für diese Bauweise (Fig. 5, rechts). Für die Druckbehälterherstellung ist diese Technologie nach aktuellem Stand noch kein etabliertes Fertigungsverfahren.

Unabhängig von der Konzeptvariante ergeben sich nach aktuellem Stand der Regulatorik Herausforderungen für die Zertifizierung der Tanksysteme. Die aktuelle Zulassungsvorschrift R134 [14] geht momentan noch von konventionellen zylindrischen Einzelbehältern mit vergleichsweise großen Durchmessern aus und berücksichtigt die vorgestellten Bauweisen nicht. Darin festgelegte Tests erlauben stellenweise per Definition keine Übertragung auf die vorgestellten Konzepte, da die Außengeometrie zu stark von den konventionellen zylindrischen Einzelbehältern abweicht. Bauraumoptimierte Bauweisen für Wasserstoffspeichersysteme erfordern also eine adäquate Weiterentwicklung der Regulatorik für deren Einsatz.

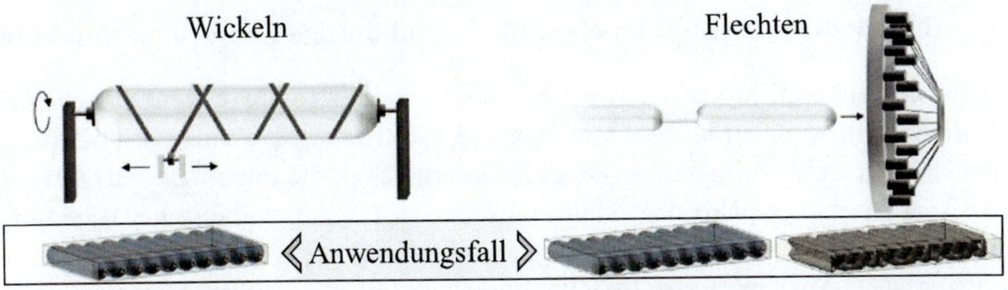

Fig. 5. Wickeln (links) und Flechten (rechts) als Fertigungsverfahren für Druckbehälter

Ein weiterer sicherheitsrelevanter und herausfordernder Aspekt für die Zertifizierung zukünftiger Druckbehälter ist die Wasserstoffpermeation, die anhand einer Leckagerate gemessen werden kann. Nach [15] erfolgt der Berechnung der Leckagerate Q in Abhängigkeit der temperaturabhängigen Permeationseigenschaften des Linerwerkstoffs P(T), dem Innendruck p, der permeationswirksamen Oberfläche A sowie der Dicke des Liners:

$$Q = P(T) \cdot \Delta p \cdot \frac{A}{t} \tag{1}$$

In [14] ist dafür ein Maximalwert pro 1 Liter Wasserstoffspeichervolumen definiert. Das bedeutet, dass eine höhere permeationswirksame Oberfläche bei gleichem Speichervolumen zu einer höheren Leckage führt. Dies ist bei zylindrischen Druckbehältern der Fall, wenn beispielsweise bei gleichbleibender Länge der Durchmesser reduziert wird. Die Einhaltung der Grenzwerte wird über die Auslegung des Liners gesteuert. Standardwerkstoffe für Liner in aktuellen Druckbehältern sind bspw. Polyamid (PA) oder High-Density Polyethylen (HDPE). In [15] wurden die Permeationseigenschaften dieser Werkstoffe in Abhängigkeit der Temperatur näher untersucht. Basierend auf diesen Permeationsergebnissen ist in Fig. 6 diejenige Linerdicke dargestellt, die zur Einhaltung der in [14] definierten Leckagerate erforderlich ist. Vor allem für Durchmesser kleiner 100 mm sind hohe Wandstärken erforderlich, die allerdings aufgrund ihres erhöhten Platzbedarfs zu einer reduzierten Wasserstoffspeichermenge führen und damit die Fahrzeugreichweite je Tankfüllung senken. Für zukünftige zylindrische Druckbehälter ist deshalb die Entwicklung neuer Werkstoffe als Permeationsbarrieren für den gespeicherten Wasserstoff zwingend erforderlich.

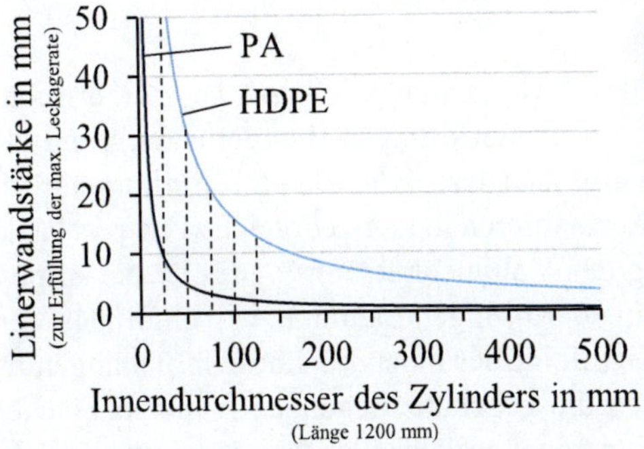

Durchmesser in mm	Linerstärke PA in mm	Linerstärke HDPE in mm
125	1,9	12,7
100	2,4	15,9
75	3,2	21,0
50	4,7	31,4
25	9,4	62,4

Fig. 6. Notwendige Linerwandstärken für PA und HDPE in Abhängigkeit des Durchmessers

2.3 Abschätzung des Einsatzpotenzials für zukünftige Fahrzeugarchitekturen

Ein entscheidendes Kaufkriterium für Kunden von Elektrofahrzeugen ist neben dem Preis auch die maximale Fahrzeugreichweite je Tankfüllung. Deshalb wird beispielhaft für das differenzielle zylindrische Druckbehälterkonzept die maximale nutzbare Wasserstoffspeichermenge analysiert. Dafür wird zunächst ein hypothetischer Bauraum im Fahrzeugunterboden mit den Abmessungen 1600 mm x 1300 mm x 130 mm definiert. Die analytische Auslegung der Behälteraußenwand basiert auf der Kesselformel und wurde mit realen Tankdaten validiert. Die Auslegung des Liners erfolgt entsprechend der oben gezeigten Randbedingungen zur Permeation. In Fig. 7 ist die Abschätzung der speicherbaren Wasserstoffmenge in Form des Volumennutzungsgrads dargestellt. Dieser beschreibt das Verhältnis des Volumens, das der nutzbare gasförmige Wasserstoff unter 700 bar Druck einnimmt, zum Bauraumvolumen, in das das Tanksystem integriert wird. Es zeigt sich, dass die Zylinder den zur Verfügung stehenden quaderförmigen Bauraum nicht optimal ausnutzen können. Dennoch ergibt sich für den Volumennutzungsgrad von 50,5 % eine nutzbare Wasserstoffspeichermenge von ca. 5,2 kg. Bei einem Durchschnittsverbrauch nach WLTP-Norm von ca. 1 kg H2 pro 100 km (vgl. [20]) resultiert eine akzeptable Fahrzeugreichweite von ca. 520 km. Im Detail ist zu prüfen, ob für ein integrales Druckbehälterkonzept eine noch höhere Reichweite aufgrund einer günstigeren Bauraumausnutzung möglich ist.

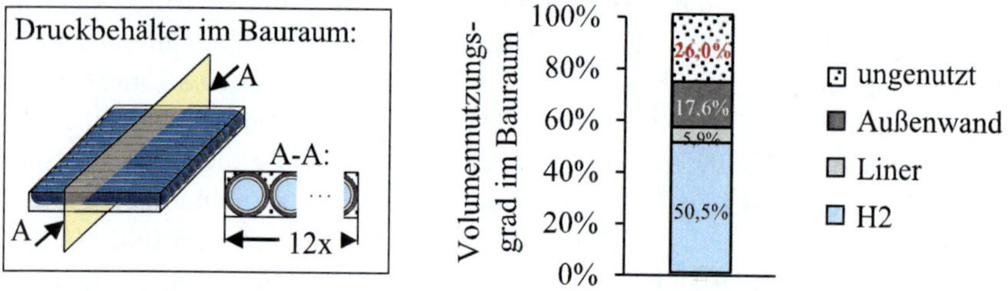

Fig. 7. Potenzialanalyse für die Speicherung von gasförmigem Wasserstoff unter 700 bar in zylindrischen Druckbehältern in einem quaderförmigen Bauraum

3 Quaderförmige Druckbehälter mit Zugstreben

3.1 Konzeptvarianten

Die durchgeführte Potenzialanalyse für den Volumennutzungsgrad der zylindrischen Druckbehälter zeigt, dass ca. ein Viertel des quaderförmigen Bauraums aufgrund des runden Behälterquerschnitts ungenutzt bleibt. Das hat einen direkten Einfluss auf die speicherbare Wasserstoffmenge sowie die resultierende Fahrzeugreichweite pro Tankfüllung. Eine Möglichkeit zur Steigerung des Volumennutzungsgrades ist die Anpassung der Druckbehältergeometrie an den quaderförmigen Bauraum. Um ein frühes Versagen oder ungleichmäßige Verformungen infolge der Innendruckbeanspruchungen für solche Geometrien zu verhindern, müssen die gegenüberliegenden planaren Flächen eines quaderförmigen Behälters durch Zugstreben miteinander verbunden werden. Im

Vergleich zum Lösungsansatz mit zylindrischen Druckbehältern weist dieses Konzept einen deutlich geringeren Technologiereifegrad auf. Mögliche konstruktive Umsetzungen wurden zwar in Patentschriften sowie Offenlegungsschriften zu deren Anmeldung aufgezeigt, jedoch nie bis zu einem annähernd serienreifen Produkt entwickelt (Fig. 8). Eine besondere Herausforderung stellt dabei die Anbindungsart der Zugstreben an die Druckbehälterwand dar, da einerseits die Kraftübertragung gewährleistet sein muss und andererseits keine Leckagen in diesem Bereich auftreten dürfen.

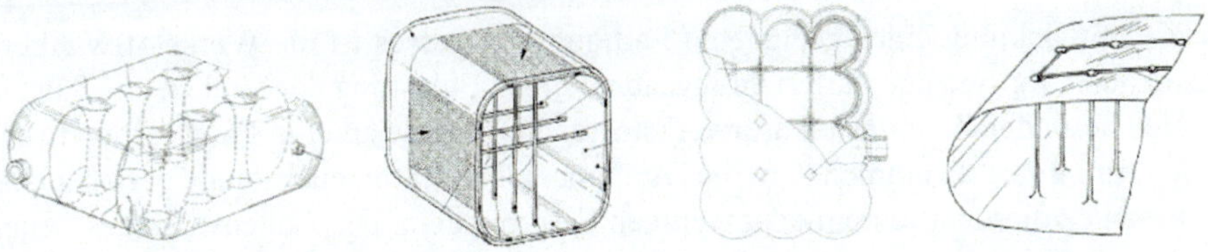

Fig. 8. Konstruktive Umsetzungsvorschläge a) [16], b) [17], c) [18] und d) [19] für Druckbehälter mit Zugverstrebungen im Behälterinneren.

3.2 Außengeometrie des Behälters und Grobauslegung der Zugstreben

Für einen perfekt quaderförmigen Druckbehälter mit sechs Außenflächen müssten Zugstreben in drei Richtungen eingebracht werden, um für die jeweils gegenüberliegenden Flächen ein Versagen oder ungleichmäßige Verformungen zu verhindern. Da die Höhe relativ zur Breite und Länge des zugrunde liegenden Bauraums gering ist, wird hier lediglich die Ober- und Unterseite des Behälters mit Streben verstärkt. Für eine belastungsgerechte Auslegung werden die Ecken und Seitenflächen abgerundet, so dass sich diese aus Viertelkugelschalen und Zylinderhalbschalen zusammensetzen (Fig. 9).

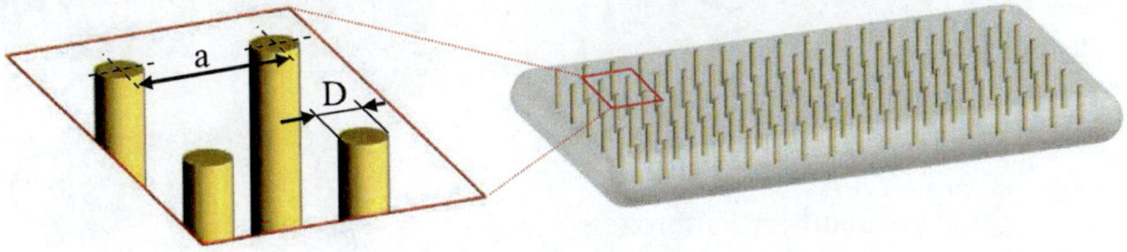

Fig. 9. Außenkontur des Druckbehälters mit schematischer Darstellung der Strebenanordnung

Eine grobe analytische Auslegung der einzelnen Zugstreben erfolgt anhand eines Kräftegleichgewichts in einer quadratischen Einheitszelle. Dabei wird jeder Strebe diejenige Fläche der Außenwand gegenübergestellt, die von ihr gehalten werden muss, um der Beanspruchung durch den Innendruck entgegenzuwirken. Entsprechend lässt sich in (2) der Abstand a zwischen den Zugstreben abhängig von Auslegungsdruck p_A, Durchmesser D und Zugfestigkeit σ_z der Strebe bestimmen.

$$a = 0{,}5 \cdot D \cdot \sqrt{\pi \cdot \left(1 + \frac{\sigma_Z}{p_A}\right)} \tag{2}$$

Generell gilt, dass größere Durchmesser größere Abstände zwischen den Streben erlauben und damit deren Anzahl geringer ausfällt. Allerdings ist zu beachten, dass größere Strebenabstände zu einer ungleichmäßigeren Aufteilung der Krafteinleitung und damit zu höheren Schubbeanspruchungen in der Außenwand führen. Aus Auslegungssicht ist deshalb ein geringer Abstand von Vorteil.

3.3 Fertigungsverfahren

Für die Entwicklung eines geeigneten Fertigungsprozesses ist die Werkstoffwahl entscheidend. Aufgrund der stark richtungsabhängigen Belastung der Streben auf Zug eignen sich besonders hochfeste Faserwerkstoffe, die zum Beispiel in Faserverbundwerkstoffen zum Einsatz kommen. Um die Durchmesser und Strebenabstände gering zu halten, bestehen diese aus einem oder wenigen Faserbündeln. Die Außenwand des Druckbehälters besteht aus carbonfaserverstärktem Kunststoff, hergestellt im Wickelverfahren. Zur form- und stoffschlüssigen Fixierung der Streben an der Druckbehälterwand durchdringen diese die Wandung und werden an der Außenseite umgelenkt. Die Einbringung der Fasern erfolgt in einem Tuftprozess, der an der Behälteroberseite einem Nähstich und an der Unterseite einem Tuftstich gleicht. Als Werkstoff für die Streben kommen aufgrund ihrer Robustheit in der Verarbeitung Aramidfasern zum Einsatz. Die Validierung des Fertigungskonzepts kann anhand erster Baumuster gezeigt werden (Fig. 10). Die Integration des Liners stellt eine besondere Herausforderung dar und ist dabei noch nicht berücksichtigt – mögliche Lösungen könnten ein nachträgliches Einbringen im Rotationsgussverfahren oder ein Konzept ohne Liner sein.

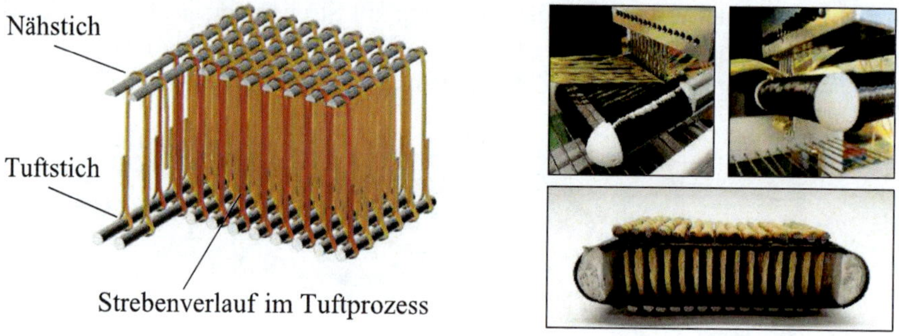

Fig. 10. Nutzung eines Tuftprozesses (links) für die Einbringung der Zugstreben in den Druckbehälter (rechts)

Erste Festigkeitsuntersuchungen erfolgen für das Konzept auf Couponebene im Bereich der Verankerung der Zugstreben an der Druckbehälterwand. Für mechanische Tests werden dafür entsprechende Proben aufgebaut, die eine Strebenumlenkung an einer Platte aus carbonfaserverstärktem Kunststoff abbilden. Für den Näh- und Tuftstich konnte so für einen ausgewählten Aramidfaserwerkstoff jeweils eine Zugfestigkeit von ca. 2000 MPa im verarbeiteten Zustand nachgewiesen werden (Fig. 11). Bei einem Strebendurchmesser von ca. 1 mm und eine Auslegungsdruck p_A von 180 MPa ergibt sich ein Strebenabstand von ca. 3 mm.

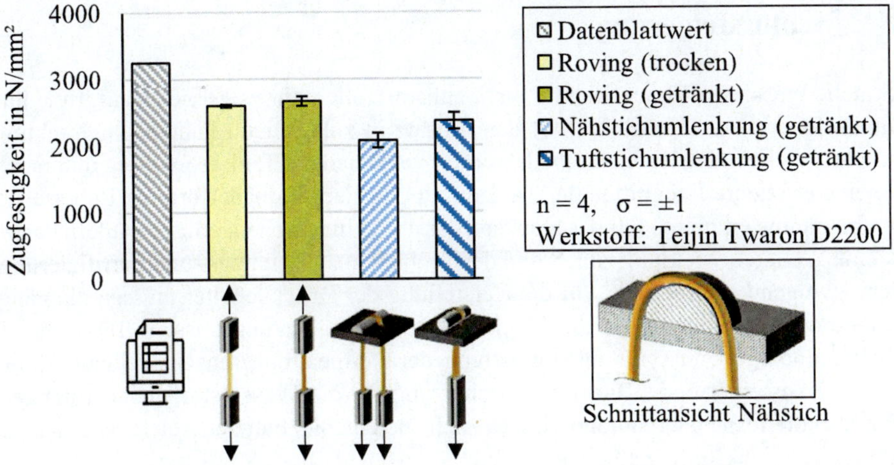

Fig. 11. Untersuchung der Zugfestigkeit einer Teijin Twaron D2200 Aramidfaser in unterschiedlichen Verarbeitungsstadien

3.4 Abschätzung des Einsatzpotenzials für zukünftige Fahrzeugarchitekturen

Analog zu den zylindrischen Tanks erfolgt auch für die zugverstrebten Druckbehälter eine Abschätzung des Einsatzpotenzials anhand der maximalen Fahrzeugreichweite je Tankfüllung. Für die analytische Untersuchung werden die gleichen Bauraumabmessungen vorausgesetzt. Anhand der Erkenntnisse zum Fertigungsverfahren und der Auslegung der Zugstreben wurde ein Modell aufgebaut, um den Volumennutzungsgrad abschätzen zu können. Die Auslegung der Außenwand erfolgt analog zu den zylindrischen Druckbehältern anhand der Kesselformel, die Linerdicke wird so eingestellt, dass die Permeationsanforderungen erfüllt werden. Aus Gründen einer günstigen Modularisierung wird der Bauraum mit vier gleich großen Behältern bestückt. Im Vergleich zum Konzept mit zylindrischen Druckbehältern bleiben nur ca. 9 % des Bauraums ungenutzt und 63,4 % des Volumens wird durch den Wasserstoff eingenommen, woraus eine nutzbare Speichermenge von ca. 6,5 kg resultiert (Fig. 12). Bei einem Durchschnittsverbrauch nach WLTP-Norm von ca. 1 kg H2 pro 100 km ergibt sich daraus eine Fahrzeugreichweite von ca. 650 km je Tankfüllung.

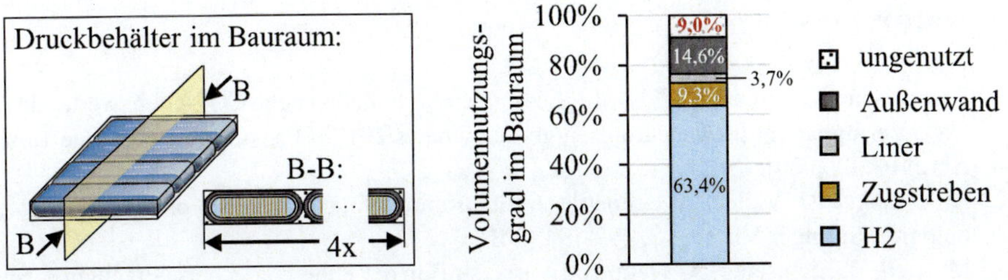

Fig. 12. Potenzialanalyse für die Speicherung von gasförmigem Wasserstoff bei 700 bar in flach ausgebildeten Druckbehältern mit Zugstreben in einem quaderförmigen Bauraum

4 Schlussfolgerung

Aktuelle Wasserstoffdruckbehälter für Brennstoffzellenfahrzeuge bieten aufgrund ihrer Dimensionen kaum Flexibilität für den Einsatz in zukünftigen Fahrzeugarchitekturen mit elektrischen Antrieben. Deshalb werden verschiedene Tankgeometrien untersucht, die eine effiziente Integration der Tanksysteme in flache, quaderförmige Batteriespeicherbauräume erlauben. Für das Konzept, das auf zylindrischen Druckbehältern basiert, ist eine Weiterentwicklung zum Stand der Technik aus Fertigungs- und Zertifizierungssicht zwingend erforderlich. Für die Herstellung der Druckbehälter müssen die gängigen Fertigungsverfahren an die kleineren Druckbehälter angepasst werden und die Linertechnologie muss für die Einhaltung der Permeationsgrenzen weiter optimiert werden. Aus einer ersten Potenzialabschätzung für die Wasserstoffspeichermenge in einem Batteriespeicherbauraum lassen sich akzeptable Fahrzeugreichweiten größer 500 km pro Tankfüllung ableiten.

Parallel wird der Ansatz eines quaderförmigen Druckbehälters verfolgt, der im Inneren mit Zugstreben verstärkt ist, um eine ungleichmäßige Verformung des Wasserstoffspeichers unter Druck zu verhindern. Die zentrale Herausforderung in der Umsetzung dieses Konzepts stellt die Anbindung der Zugstreben an die Druckbehälterwand dar. In diesen Bereichen muss einerseits die Kraftübertragung sichergestellt werden, andererseits muss die Dichtheit gewährleistet werden und es dürfen auch infolge geringfügiger Ausdehnungen durch die Druckbelastung keine Leckagen entstehen. Das entwickelte Fertigungsverfahren muss weiter untersucht werden, um den Technologiereifegrad aus dem Forschungsstadium zu heben. Ersten Analysen zufolge birgt das Konzept im Vergleich zu den zylindrischen Druckbehältern ein um ca. 25 % höheres Reichweitenpotenzial durch eine effizientere Bauraumausnutzung.

Eine Bestätigung der gezeigten Potenzialabschätzungen ist noch ausstehend. Erste Tests zur Ermittlung der Berstdrücke für die Konzepte sollten im nächsten Schritt Aufschluss zur Realisierbarkeit der Konzepte geben.

Danksagung. Die aufgezeigten Untersuchungen wurden unter anderem im Rahmen des Forschungsprojekts BRYSON durchgeführt. Das noch laufende Projekt wird gefördert durch das Bundesministerium für Wirtschaft und Energie.

Literatur

1. Europäische Kommission, „EU-Mitgliedstaaten beschließen neue CO2-Grenzwerte für Autos". Verfügbar: https://ec.europa.eu/germany/news/20190415-co2-grenzwerte_de (letzter Abruf 03.09.2020).
2. A. Kampker, D. Vallée, A. Schnettler, „Elektromobilität: Grundlagen einer Zukunftstechnologie", Springer Vieweg, S. 192-194 (2018).
3. M. Klell, H. Eichlseder, A. Trattner, „Wasserstoff in der Fahrzeugtechnik: Erzeugung, Speicherung, Anwendung", Springer Vieweg, S. 126-127, 180-182 (2018).
4. K. Sasaki, H. Li, A. Hayashi, J. Yamabe, T. Ogura, S. M. Lyth, „Hydrogen Energy Engineering: A Japanese Perspective", Springer, S. 465 (2016).

5. Bayerische Motorenwerke Aktiengesellschaft.
6. J. Töpler, J. Lehmann, „Wasserstoff und Brennstoffzelle: Technologien und Marktperspektiven“, Springer Vieweg, S. 92 (2017).
7. AVK – Industrievereinigung Verstärkte Kunststoffe e.V., „Handbuch Faserverbundwerkstoffe/ Composites: Grundlagen, Verarbeitung, Anwendungen“, Springer Vieweg, S. 362 (2014).
8. Mercedes-Benz AG, „Mercedes-Benz GLC F-Cell“. Verfügbar: https://www.mercedes-benz.de/passengercars/mercedes-benz-cars/models/glc/glc-f-cell/der-neue-glc-f-cell/film.module.html (letzter Abruf 03.09.2020).
9. Schutzrecht DE 102018116090 A1, „Hochdruck Behältereinheit“, Toyota Motor Company (2019).
10. Schutzrecht DE 102013002944 A1, „Vorrichtung zum Speichern von Gas unter hohem Druck“, Daimler AG (2014).
11. Schutzrecht DE 102017214606 A1, „Brennstofftank und Fahrzeug“, Audi AG (2019).
12. Schutzrecht US 9217538 B2, „Conformable Natural Gas Storage“, Other Lab LLC (2015).
13. E. Bigelow, M. Lewis, „Conformable Hydrogen Storage Pressure Vessels“, Technical Report, US Department of Energy (DOE), Office of Energy Efficiency and Renewable Energy (2018).
14. Regelung Nr. 134 der Wirtschaftskommission der Vereinten Nationen für Europa (UNECE), „Einheitliche Bestimmungen für die Genehmigung von Kraftfahrzeugen und Kraftfahrzeugbauteilen hinsichtlich der sicherheitsrelevanten Eigenschaften von mit Wasserstoff und Brennstoffzellen betriebenen Fahrzeugen (HFCV)“ [2019/795]: R 134, (2019).
15. J. Humpenöder, „Gaspermeation von Faserverbunden mit Polymermatrices”, Dissertation, Universität Karlsruhe (TH), S. 5, 74 (1997).
16. Schutzrecht WO 0218836 A1, „Device for storing pressurized fluid and/or reactors generating pressurized fluids, in particular for motor vehicles, Renault (2002).
17. Schutzrecht DE 102009057170 A1, „Druckbehälter zur Speicherung von fluiden Medien, insbesondere zum Einbau in ein Fahrzeug“, Volkswagen AG (2011).
18. Schutzrecht DE 19725369 A1, „Nichtzylindrischer Verbundstoffdruckbehälter“, Braune M. (1998).
19. Schutzrecht DE 19749950 C2, „Behälter zum Speichern von Druckgas“, Mannesmann AG (1999).
20. Hyundai Motor Deutschland GmbH, „Der Hyundai NEXO“. Verfügbar: https://www.hyundai.de/downloads/modelle/preisliste/hyundai-nexo-preisliste-juni-2020.pdf/ (letzter Abruf: 03.09.2020).

Bauraumoptimierter Wasserstoff Tank mit innerer Zugverstrebung

Alexander Horoschenkoff[1], Martin Huber[1], Alexander Hupfeld[1[0000-0001-9979-7643]],

[1] Hochschule für angewandte Wissenschaften München,
Dachauer Str.98b München, Deutschland

alexander.horoschenkoff@hm.edu

Kurzfassung. Im Rahmen der vom Bundeswirtschaftsministerium geförderten BRYSON Projektgruppe (BauRaumeffiziente HYdrogenSpeicher Optimierter Nutzbarkeit) wird an der Hochschule München an einem zugverstreben Wasserstofftank für PKW mit Brennstoffzellenantrieb geforscht, der effizient in die Unterbodenstruktur integriert werden kann. Die Untersuchungen zeigen die grundlegende Machbarkeit eines flachen, kubischen Drucktanks mit innerer Zugverstrebung. Wesentliche Aspekte, wie das Tragverhalten bei hoher Strebenanzahl, Dichtigkeit und Fertigbarkeit konnten untersucht und analysiert werden.

Stichwörter: FCEV, Wasserstoffdrucktank, Tankintegration

1 Problemstellung, Bauweise und Dichtigkeit

Die Integration der Wasserstofftanks bei heutigen Brennstoffzellenfahrzeugen erfolgt üblicherweise in einem oder mehreren zylindrischen carbonfaserverstärkten Druckbehältern, die im Antriebswellentunnel und im Heckbereich des Fahrzeugs angeordnet sind. Dies führt in der Regel zu einer Verringerung des Kofferraumvolumens und der Fahrdynamik.
Entgegen dem aktuellen Stand der Technik werden die Tanks mit innerer Zugverstrebung quaderförmig ausgebildet, um sie effizient in die flachen Bauräume der Unterbodenstruktur des Fahrzeugs integrieren zu können. Somit können Baureihensynergien mit zukünftigen batteriebetriebenen Elektrofahrzeugen genutzt werden. Die Randbedingungen für den schnellbetankungstauglichen Tank nach SAE 2601 Standard sind ein Betriebsdruck von 875 bar und ein Berstdruck von 1600 bar bei einem nutzbaren Wasserstofffassungsvermögen von ca. 5,6 kg.

Die entwickelte Bauweise sieht vor, dass CFK-Umfangswicklungen die Druckkräfte der Stirnfläche aufnehmen und die Zugstreben im Tankinneren die Aufweitung des Tanks in Dickenrichtung verhindern. Die Strebenkräfte werden über eine Schlaufenkonstruktion und eine metallische Umlenkung aufgenommen (siehe Abb. 1). Analytische Berechnungen und stochastische Berechnungen zum Einfluss von Fertigungstoleranzen auf die Tragfähigkeit des Strebenverbunds ergeben, dass bei einem notwendigen

S. Pfingstl et al. (eds.), *Proceedings of the Munich Symposium on Lightweight Design 2020*, https://doi.org/10.1007/978-3-662-63143-0_9

Strebenabstand von ca. 5mm Aramidfasern mit geringem E-Modul als Strebenmaterial besser geeignet sind als Carbonfaser-Streben (siehe Abb. 2).

Abb. 1. Baumuster zum Tank mit innerer Zugverstrebung.

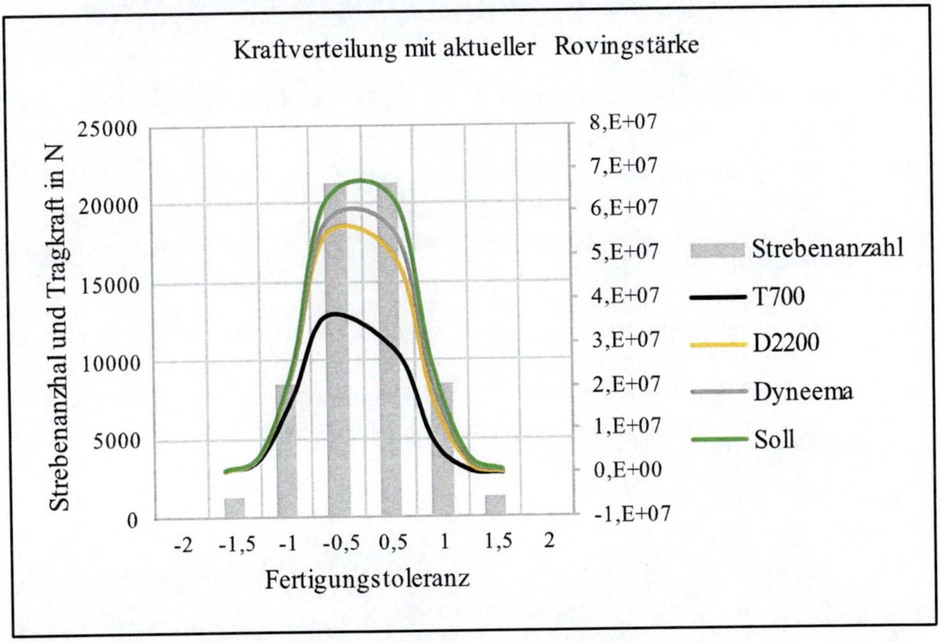

Abb. 2. Stochastische Untersuchung zum Einfluss von Fertigungstoleranzen auf die Tragfähigkeit des Strebenverbunds bei unterschiedlichen Strebenmaterialien.

Eine wesentliche Problematik des Tanks mit Zugverstrebung ist die Sicherstellung der Dichtigkeit über eine Linerschicht. Hierzu wurden analytische Berechnungen zur Bestimmung der Schubspannungen zwischen Liner und Strebe durchgeführt und experimentelle Versuche zur Überprüfung der Dichtigkeit der Strebendurchführung. Die experimentellen Versuche sehen vor, dass einzelne Strebendurchführungen zyklisch belastet werden und anschließend die Dichtigkeit mit Hilfe eines Prüfstands, der mit Formiergas betrieben wird, überprüft wird. Anhand von Schliffen konnte die Entstehung intralaminarer Risse in den CFK-Lagen nachgewiesen werden (siehe Abb. 3).

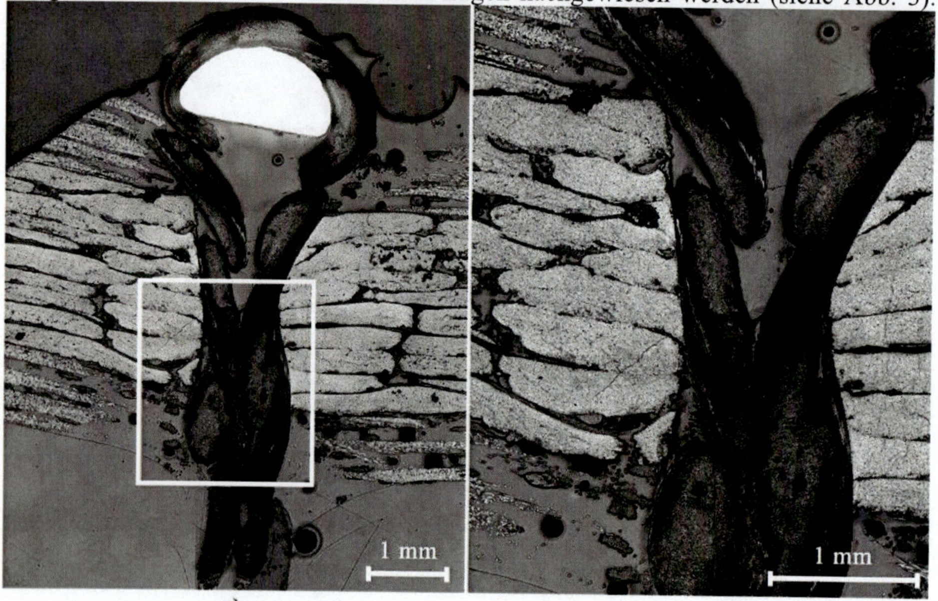

a) b)

Abb. 3. a) Schliffbild Einbettung der Schlaufenumlenkung b) Intralaminare Risse Tankwand.

2 Fertigungskonzept

Studien zum Fertigungsprozess zeigen, dass mit dem Tuftprozess effizient Streben in die CFK-Wickelstruktur eingebracht werden können. Durch die Verwendung einer großen Anzahl von Tuftnadeln entsteht ein für den Tuftprozess charakteristischer Schlaufentunnel, der zum Einlegen der Umlenkung genutzt werden kann (siehe Abb. 4).

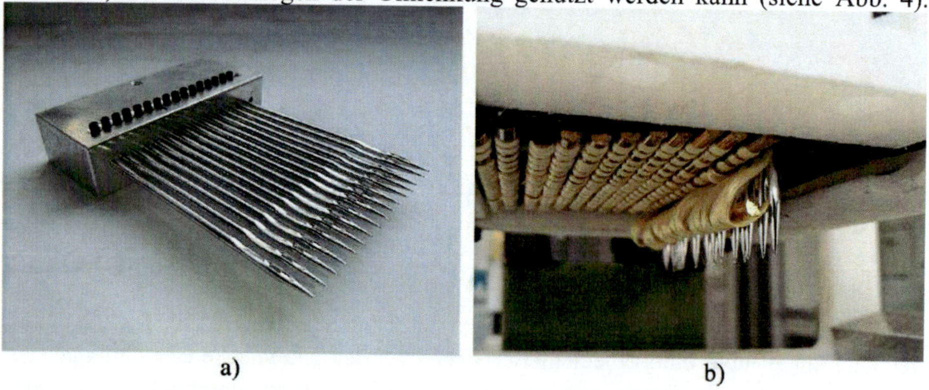

a) b)

Abb. 4. a) Tuftkopf mit 16 Nadeln b) Einbringen der Schlaufenumlenkung in den sich beim Tuftprozess bildenden Schlingentunnel

Als Linerwerkstoffe werden Epoxidharze und Gusspolyamide untersucht, die über einen Rotationsformprozess eingebracht werden. Mit prototypischen Anlagen wurden erste Baumuster hergestellt und strukturmechanisch charakterisiert.

3 Zusammenfassung

Mit Hilfe der Untersuchungen konnte eine grundlegende Bauweise für einen kubischen, flachen Drucktanks mit innerer Zugverstrebung aus Faserverbundwerkstoffen entwickelt werden. Wesentliche Aspekte, wie das Tragverhalten bei hoher Streben Anzahl, Dichtigkeit und Fertigbarkeit konnten analysiert und experimentell untersucht worden. Auf den Ergebnissen aufbauend können nun entsprechende Parameterstudien gezielt durchgeführt werden oder aber Alternativen aufgezeigt werden.

Literaturverzeichnis

1. Lynch, F., Eastridge, E., Everett, G.: Pressure Vessels and Method of Fabrication. Offenlegungsschrift zur Patentanmeldung, US 20200318789 A1, (2020).
2. UN ECE R134, Uniform provisions concerning the approval of motor vehicles and their components with regard to the safety-related performance of hydrogenfuelled vehicles (HFCV). (2015).
3. Eyerer, P., Hirth, T., Elsner, P.: Polymer Engineering. Springer Verlag, Berlin/Heidelberg (2008).
4. Cherif, C.: Textile Werkstoffe für den Leichtbau. Springer Verlag, Berlin/Heidelberg (2020).

Gefördert durch:

Bundesministerium für Wirtschaft und Energie

aufgrund eines Beschlusses des Deutschen Bundestages

Innovative design and manufacturing techniques for fiber reinforced plastic components

Klemens Rother[1][0000-0003-9643-4967] and Stephen W. Tsai[2]

[1] Munich University of Applied Sciences, 80335 Munich, Germany
[2] Stanford University, Stanford, CA 94305-4035, USA
klemens.rother@hm.edu

Abstract. This paper covers new and paradigm changing findings in analysis, optimization, design and manufacturing of endless fiber composites as developed by an international group of experts in research and industry inspired and driven by Prof. em. Stephen Tsai. This was presented at the 2019 Munich Lightweight Symposium and is added to the 2020 proceedings as an introduction of the application as presented by Neuhäusler in this year's symposium [7]. The fundamental finding used for the new concepts is the normalizing property of the trace of the stiffness matrix of a unidirectional ply as an invariant material property of plies and laminates. This as a basis enables the use of variable ply angles, a so-called Double-Double architecture of sub-laminates and many more. Much simpler stackings, less design rules to follow, increased strength of laminates, thinner minimum gages, simplified optimization and easier and cost-efficient manufacturing can be obtained.

Keywords: Trace, Double-Double, Tsai's Modulus

1 Introduction

Traditional quad-axial sub-laminates to create complex structures, so called "legacy quads", face many challenges for design and manufacturing. By limiting to $[0_p/\pm45_q/90_r]_S$ sub-laminates only discrete basic laminates can be chosen to find optima due to too many possible permutations in a sub-laminate consisting of 6, 8 or 10 plies. (i) high effort, (ii) experimental skill to collect all material data for stiffness and strength of unidirectional plies, (iii) many design rules to follow to create more complex stackings to avoid coupled bending, warping and shear deformations, (iv) a problematic layup and fiber unravel (especially occurring in corner regions of [+45/-45]-plies), (v) high cost for design and manufacturing are just some of the obstacles for easy design, quick optimization and simpler manufacturing of components by legacy quads. Also, stackings of such basic laminates become unnecessary thick, face lower strength and are prone to delamination.

As a fundamental finding by Tsai and Melo in 2014 [1] a novel invariant-based approach to describe the stiffness of composite plies and laminates that takes the trace of the plane stress stiffness matrix of a unidirectional ply as a material property was presented (tr**Q** or simply referred to as *trace*). Since tr**Q** is an invariant with respect to coordinate transformations the trace of the thickness-normalized in-plane ($\mathbf{A}^*$) and

S. Pfingstl et al. (eds.), *Proceedings of the Munich Symposium on Lightweight Design 2020*, https://doi.org/10.1007/978-3-662-63143-0_10

flexural ($\mathbf{D}^*$) stiffness matrices of laminates are also the same: tr$\mathbf{A}^*$= tr$\mathbf{D}^*$= tr$\mathbf{Q}$. This idea has proven to be a "door-opener" for more paradigm-changing solutions as a toolbox, like (i) a simplified Unit-Circle failure criterion, (ii) Master- and Universal Ply constants for the geometrical layout of single plies as well as stacked laminates, (iii) the possibility of easy homogenization of laminates to avoid complex rules for design and manufacturing, (iv) field-equations for arbitrary variable ply angles, and (v) a straight forward *lamsearch* approach which minimizes numerical efforts significantly. (vi) using a Double-Double architecture of sub-laminates further simplifies the stacking sequences of laminated structures and enables an efficient sizing method to be used. Double-Double might be further extended to (vii) skin-grid structures to activate even more lightweight potentials. Those concepts will be briefly described in this paper and can be found in [1]-[4][7].

2 Trace (Tsai's modulus[1]) on engineering constants

For the classical lamination theory CLT, many engineering constants are needed to represent stiffness and strength of a uniaxial ply. On top, each laminate needs to be tested individually by coupon testing to supply 4 necessary constants for stiffness as can be seen in equation (1) (ν_y might be calculated using the other constants).

$$\mathbf{Q}=\begin{bmatrix} \frac{E_x}{1-\nu_x\nu_y} & \frac{\upsilon_y E_x}{1-\nu_x\nu_y} & 0 \\ \frac{\nu_x E_y}{1-\nu_x\nu_y} & \frac{E_y}{1-\nu_x\nu_y} & 0 \\ 0 & 0 & G_{xy} \end{bmatrix} = \begin{bmatrix} Q_{xx} & Q_{xy} & 0 \\ Q_{xy} & Q_{yy} & 0 \\ 0 & 0 & Q_{ss} \end{bmatrix} \tag{1}$$

The invariant theory based on trace reduces the need for testing those constants for stiffness to just one: trace. Tr$\mathbf{Q}$ is a scalar stiffness value derived from a unidirectional on- or off-axis ply or even a more complex laminate of the material chosen. Tr$\mathbf{Q}$ and its normalization properties as shown in the next chapter is the fundamental and only material property needed to be tested to obtain the stiffness properties needed. Some of the engineering constants like the shear modulus G_{xy} are difficult to measure and thus prone to error but with tr$\mathbf{Q}$ they can be estimated with sufficient accuracy as will be shown in the next chapter.

---

1 Comment by corresponding author: following these new design principles, a number of researchers and experts have taken those ideas for applications and further developments in many aspects. See for example all authors of [4] or especially [6] who are involved. This group suggested in [6] to name the fundamental property, tr$\mathbf{Q}$, in honor to the fundamental work of Stephen W. Tsai as "Tsai's Modulus".

Using the simplified Unit-Circle failure criterion in addition, only two more constants, X and X' for a unidirectional ply are needed for strength evaluation as will be discussed later.

As a consequence, this leads to massive reduction of material testing effort and cost (simple uniaxial tension and compression test remaining, no more difficult parameters to measure). Trace and X, X' are sufficient for simpler and much faster optimization of laminates covering variable ply angles, stacking, local thickness and material selection.

One material parameter for stiffness (tr**Q**) added by *Universal Ply* or *Master Ply Constants* which are dependent on the material group only, finally and fully describe the stiffness and engineering constants of any ply and specific material.

3 Universal and Master Ply Constants and Laminate Factors

If tr**Q** represents an accumulated stiffness and is the only material constant needed for stiffness of a ply and even a laminate, how can engineering constants or ply stiffness values Q_{ij} be extracted from tr**Q**? This can be done using so called *Universal Ply Constants* or *Master Ply Constants*. In the following Table 1 a selection of only 4 CFRP materials are listed with their ply engineering constants and on-axis ply stiffness values. Watch the material dependence of tr**Q** = $Q_{xx}+Q_{yy}+2Q_{ss}$ as listed in the last column. Because of the engineering shear values are used a factor of 2 applies to Q_{ss}.

Table 1. Ply engineering constants and on-axis ply stiffness with tr**Q**

	Ply engineering constants				On-axis ply stiffness Q_{ij}				
	E_x	E_y	G_{xy}	ν_x	Q_{xx}	Q_{yy}	Q_{xy}	Q_{ss}	*trQ*
	GPa	GPa	GPa	[-]	GPa	GPa	GPa	GPa	GPa
IM6/epoxy	203	11,2	8,4	0,32	204	11,3	3,6	8,40	232
T800/Cytec	162	9,0	5,0	0,40	163	9,1	3,6	5,00	183
AS4/3501	138	9,0	7,1	0,30	139	9,1	2,7	7,10	162
T700/2510	126	8,4	4,2	0,31	127	8,5	2,6	4,20	144
Mean of total from 15 CFRP				0.32					
CoV applied to 15 CFRP				8.7%					

Table 2 shows the trace-normalized stiffness values Q_{ij}^* and the trace-normalized engineering constants. The trace of both yields always 1.0. Arithmetic mean values for Q_{ij}^* and the normalized engineering constants are shown below in the Table 2 for a total of 15 CFRP contained in those numbers. Especially for the dominating values Q_{xx}^* and E_x^* these averaged values have only low coefficients of variation of less than 2%.

The *Universal Ply Constants* and *Master Ply Constants* from Table 1 and Table 2 apply to all on-axis CFRP. Multiplication those constants by tr**Q** yields the Q_{ij} and the elastic engineering constants of the respective material.

The constants listed are valid for CFRP only. There have been investigations on aramid and glass fiber composites as well, both resulting different constants. The results for aramid are in a similar small scatter band like CFRP. Glass has shown to be less accurate but still applicable especially for conceptual studies.

As a result, knowing tr**Q** and the master ply and universal ply constants the Q_{ij} and engineering constants can simply be back-calculated. This way tr**Q** is the one and only material constant which is needed to represent stiffness of those anisotropic composites.

Table 2. Trace normalized ply stiffness values and engineering constants

	Trace-normalized on-axis ply stiffness Q_{ij}^*					Trace-normalized eng. constants			
	Q_{xx}^*	Q_{yy}^*	Q_{xy}^*	Q_{ss}^*	*trQ**	E_x^*	E_y^*	G_{xy}^*	*tr*
	GPa	GPa	GPa	[-]	[-]	[-]	[-]	[-]	[-]
IM6/epoxy	0,879	0,049	0,016	0,036	**1,0**	0,874	0,048	0,036	**1,0**
T800/Cytec	0,895	0,050	0,020	0,027	**1,0**	0,888	0,049	0,027	**1,0**
AS4/3501	0,857	0,056	0,017	0,044	**1,0**	0,851	0,056	0,044	**1,0**
T700/2510	0,883	0,059	0,018	0,029	**1,0**	0,877	0,058	0,029	**1,0**
15 CFRP	**Master Ply for CFRP**					**Universal Ply for CFRP**			
Mean	0.885	0.052		0.031		0.880	0.052	0.031	
CoV	1.4%	9.7%		17.3%		1.4%	9.7%	17.3%	

The power of Tsai's modulus becomes even more visible with respect to off-axis properties of unitape or any laminate as seen in Fig. 1. tr$\mathbf{Q}^*$ remains constant even for any ply angle. Expanding this to more complex laminates, trace of the thickness normalized membrane stiffness $\mathbf{A}^*$ behaves the same. This means: no matter of having a unitape of any ply angle or any laminate, tr**Q** can be obtained from a single coupon test.

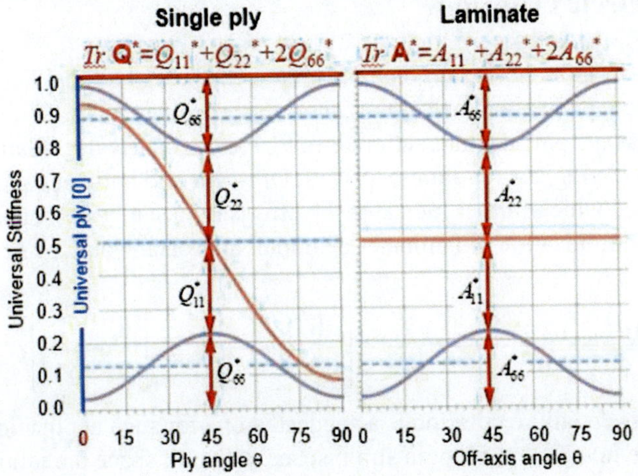

Fig. 1. Composition and traces for thickness normalized Q^* and A^*

Since tr**Q** as an invariant of the stiffness tensor **Q** is independent of the coordinate system, the number of layers, layup sequence and loading condition (in-plane or flexural) and if the material system remains the same, tr**Q** = tr**A***= tr**D***. Thus, trace is the only material constant not only to define the lamina stiffness but also the laminate stiffness as a stack of multiple plies!

For any stacking, we can define laminate related Universal Constants = *Laminate Factors* - as seen for a selection of laminates in Table 3. Those *Laminate Factors* are valid for any CFRP.

Table 3. Laminate factors for selected stacking sequences

	E_x^o/Tr	E_y^o/Tr	G_{xy}^o/Tr	ν_x^o
Universal (master) ply [0]	**0,880**	**0,052**	**0,031**	**0,320**
[0/90]	0,468	0,468	0,030	0,036
[π/*n*]	0,336	0,336	0,129	0,308
[0_7/±45/90]	0,662	0,175	0,070	0,310
[0_5/±45_2/90]	0,518	0,208	0,109	0,423
[0_2/±45/90]	0,445	0,289	0,109	0,308
[0/±45_4/90]	0,217	0,217	0,187	0,552
[0/±45]	0,370	0,155	0,161	0,734
[0/±45/0]	0,499	0,141	0,129	0,701
[0/±30]	0,510	0,074	0,129	1,220
[0/±30/0]	0,611	0,072	0,104	1,079
[±12,5]	0,764	0,053	0,066	0,913

4 Unit circle criterion

A state-of-the-art failure criterion requires more constants to be supplied besides stiffness. As can be seen in equation (2) the criterion according to Tsai-Wu for example requires 5 constants plus the interaction term F_{xy}^* and a matrix degradation factor E_m^* in case of considering last ply failure as well. Dry-cold and hot-wet conditions on top need to be tested which further increases the efforts. Especially for conceptual design and first level optimization of laminates a simpler approach would be helpful.

$$\frac{\sigma_x^2}{XX'}+\frac{2F_{xy}^*\sigma_x\sigma_y}{\sqrt{XX'YY'}}+\frac{\sigma_y^2}{YY'}+\frac{\sigma_s^2}{S^2}+\left[\frac{1}{X}-\frac{1}{X'}\right]\sigma_x+\left[\frac{1}{Y}-\frac{1}{Y'}\right]\sigma_y=1 \quad (2)$$

The *Unit-Circle* failure criterion is a suggested one for such applications which also applies the invariant concept in strain space. In strain space the failure criterion can be defined independent from the laminate layup composition because strains are linearly distributed through thickness whereas stresses are not. Failure envelopes remain the same shape independent from the number of plies involved in the laminate.

A superposition of the failure envelopes of all possible ply orientations gives a conservative inner hull of the allowable strain space for the full laminate as an invariant. Fig. 2 shows the failure envelopes for just three selected ply angles and the inner hull to be used for failure assessment. This can be expanded to any possible angle and used for any biaxial loading, called omni-strain envelope and thus is representing the failure in principal strains ε_I and ε_{II} of the laminate. Any biaxial strain of any material can be checked using this failure envelope.

Those failure envelopes are plane strain representations obtained from plane-stress failure criterions. Normalizing the failure envelope by failure strains in tension ε_X and compression $\varepsilon_{X'}$ gives the unit circle as can be seen in Fig. 3 for different materials.

Compared to Tsai-Wu failure criterion, the *Unit-Circle* offers additional safety as can be seen in the corner regions of the envelopes in Fig.3. Especially in biaxial compression the differences are higher. *Unit Circle* complies with Tsai-Wu at anchor points for uniaxial straining exactly for all cross-ply laminates (containing 0° and 90° plies).

Advantage: only tensile and compressive strength X and X' need to be provided but the use of the Unit Circle Criterion is not mandatory for the optimization. Maximum simplification might be obtained by just using X in both loading directions and neglecting X'.

Strength is checked by the strength ratio R of allowable to existing strains applied to the failure envelope in principal strain space. This metric provides the basis for an efficient linear scaling of the thickness to get the optimum laminate thickness at each location (or alternatively providing a factor on the applied loading to reach the envelope). Of course, any failure criterion, formulated in strain space, can be used.

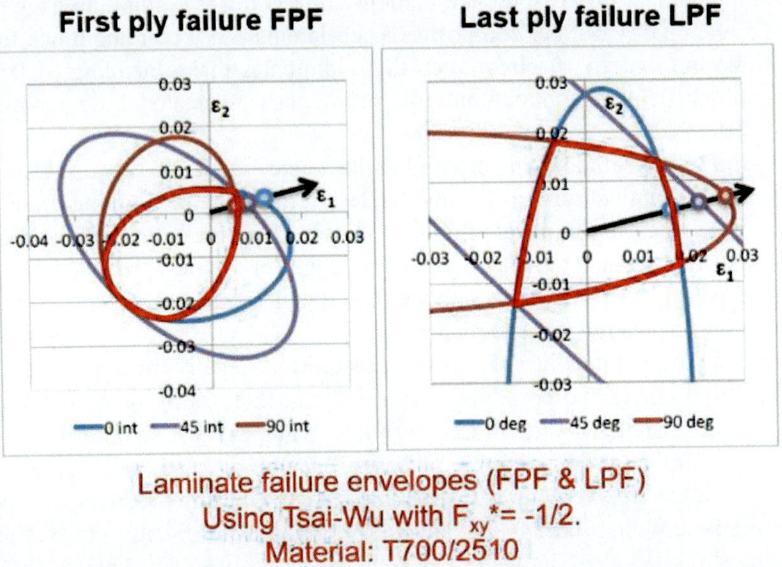

Fig. 2. Single and superposed inner strain space failure envelopes shown for three ply angles only and for first ply failure FPF (left) and last ply failure LPF (right)

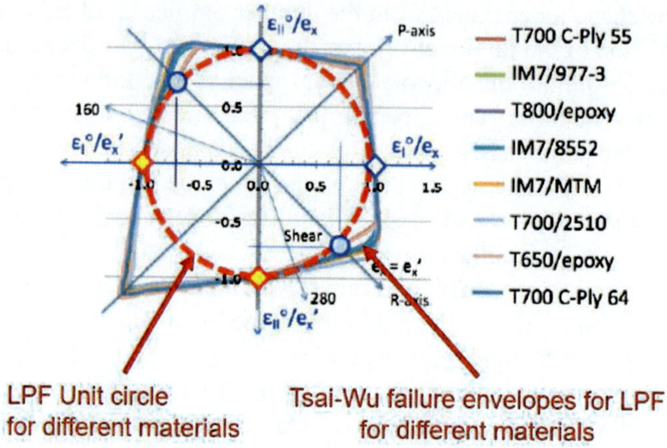

Fig. 3. Unit-Circle for different materials compared to Tsai-Wu failure envelopes, both for LPF

5 Double-double architecture for sub-laminates

5.1 Field equation for stiffness

Traditional „legacy quad" design $[0_p/\pm45_q/90_r]_S$ requires to follow many design rules like mid-plane symmetry (to avoid warpage from curing), stacking rules like the 10% rule, nesting no more than 3 plies of the same angle, balancing (for in-plane stiffness to remain orthotropic and no in-plane stretch vs. shear coupling occurs) or others. Due to an exploding number of possible permutations and to limit the minimum gage but also forming directionality of fiber composites a sub-laminate as a building block for laminates should not exceed 10 plies. Such thick laminates make blending of laminates among zones difficult and not optimal. Internal notches are created. Layup in manufacturing is difficult.

Fig. 4 shows the solutions in carpet plots for legacy quads on the left. Using up to 10 plies in a sub-laminate and following the design rules, the best among these limited to 29 possible discrete laminates might be chosen. Optimization of the stacking sequence of such laminates remains a great challenge for the designers due to the high number of possible permutations and unidirectional repeats necessary to minimize weight and directionality of the laminate.

Introducing two biaxial plies of only two positive and negative but continuous angles in total $[\pm\Phi/\pm\Psi]$ gives the possibility to describe stiffnesses by field equations as a function of those angles Φ and Ψ. Each of the two angles must be optimized to "earn its place" in a laminate which requires a "best of" search. This new architecture of sub-laminates is called Double-Double. Using thin plies even further reduces the thickness of such simple building blocks. The advantage can be seen directly in Fig.4 at right. The stiffness of quad-axial laminates can be at least closely matched by corresponding Double-Double settings. Square symmetric laminates (equal numbers of 0° and 90° plies, cross-ply ratio = 1:1) can be matched exactly by double-double laminates. In Fig.

4 one such replacement of a [0/±45/90]S having 8 plies is shown. The corresponding Double-Double requires [±22,5/±67,5] or [±67,5/±22,5], each of them requiring the thickness of just 2 regular plies in thickness by using thin plies for Double-Double. These specific laminates would require a single biaxial [±22,5] tape only but with cross ply layup 1:1 as the simplest solution. Such "black metal" solutions are not in our focus but just show the possibility of exact or close matching. The power of such a field equation makes it possible to simplify the optimization significantly.

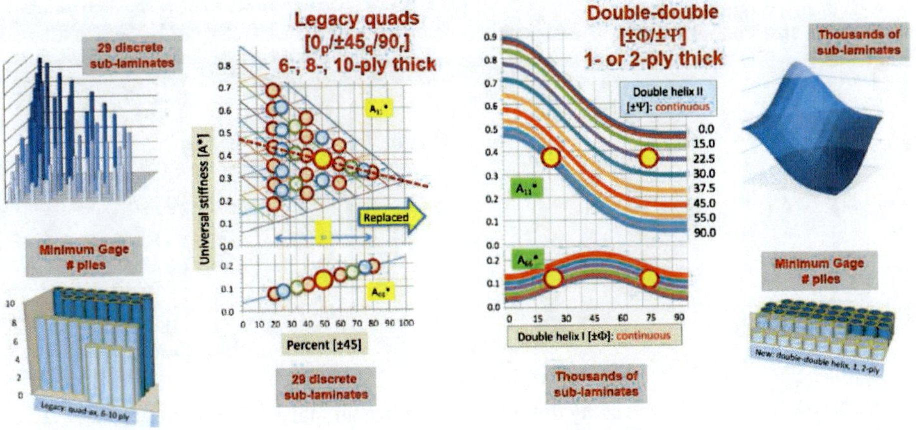

Fig. 4. Double-Double architecture of laminates $[\pm\Phi/\pm\Psi]_{rT}$ (right) compared to traditional quadaxial laminates $[0_p/\pm45_q/90_r]_S$ ("legacy quads", left) shown in carped plots

Besides the simplified layup, those thin-ply Double-Double laminates offer more advantages like increased resistance to micro-cracking (see Fig. 5, right), reduced cost by using spread tows, reduced notch and edge effects (see Fig. 5 right) and thus increased strength in general. Probably the most important aspects are (i) faster layup of complex and thick laminates as explained in the next chapter, (ii) a simpler design, (iii) an easier tapering through single ply drops and (iv) easier merge or transition between different sub-laminates in adjacent bays. Having thinner sub-laminates will reduce the minimum gage barrier that often prohibits use of composites because the resulting minimum gage is heavier and more expensive as compared to solutions using other materials (aluminum, high strength steel, ...). There are lots of manufacturing advantages by using Double-Double, especially simplifications for automatized layup, e.g. using automatic fiber placement or automatic tape placement.

5.2 Homogenization

Having thin Double-Double sub-laminates, required strength and stiffness locally is achieved by stacking multiple Double-Double sub-laminates of same angles on top of each other. Minimum gages can be achieved because the sub-laminates are minimum in thickness as a basic building block of more complex laminates. Stacking at least 5 to 8 repeats of Double-Doubles, homogenization of the total laminate might be sufficiently achieved. Homogenization means, that the **B**-matrix approaches zero and thus the design requirement of a symmetric laminate is not important any more. Layup can be done completely independent from symmetry requirements, e.g. from inside-out or outside-in. This effect brings one of the most important aspects for massive cost reduction in manufacturing of such components.

Fig. 5. Double-Double designs using thin plies don't warp and face increased strength

5.3 Quad-axial replacements

The power of Double-Double can be seen in the following example of a replacement of 11 quad-axial laminates of a fuselage as published by NASA. See Fig. 6.

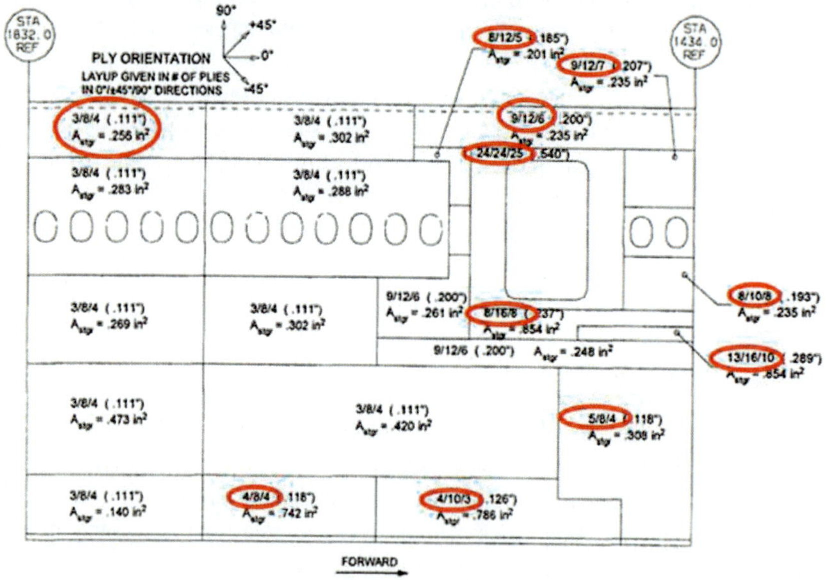

Fig. 6. Quad-axial laminates of a fuselage as published by NASA [8]

As shown in Table 4 individual replacements of these 11 laminates can be achieved by 11 Double-Double's which then might be averaged to get a single replacement for all 11 laminates by $[\pm 20/\pm 67]_{rT}$. Watch the good agreement of all the stiffness values A_{11}, A_{22}, A_{66} between the quad-axial and the Double-Double replacements but also the standard deviations of the Double-Double angles and the corresponding stiffnesses with respect to the averaged values. The scatter turns out to be less than 5% for all of those replacements measured in standard deviation. This simplified analysis of this case should demonstrate the power of such replacements.

Table 4. Replacement of 11 quad-axial laminates of a fuselage by one Double-Double $[\pm 20/\pm 67]_{rT}$

T700 C-Ply64

n [0]	n [45]	n [90]	% [0]	% [45]	% [90]	A11	A66	A22	φ	ψ	A11	A66	A22
3	8	4	20%	53%	27%	54,7	21,9	64,1	26	69	54,1	22,1	64,4
9	12	6	33%	44%	22%	69,0	19,1	55,5	16	65	69,2	18,6	56,2
8	10	8	31%	38%	31%	64,2	17,2	64,2	19	71	63,6	17,4	64,3
13	16	10	33%	41%	26%	68,0	18,1	58,5	17	66	67,8	18,6	57,8
4	8	4	25%	50%	25%	60,4	21,0	60,4	22	67	60,4	21	60,4
4	10	3	24%	59%	18%	62,2	23,9	52,7	22	62	62,4	24	52,4
8	12	5	32%	48%	20%	69,1	20,4	52,9	17	63	68,8	20,2	53,5
9	12	7	32%	43%	25%	67,3	18,8	57,9	17	66	67,8	18,6	57,8
24	24	25	33%	33%	34%	65,1	15,6	66,5	18	73	65,1	15,5	66,5
8	16	8	25%	50%	25%	60,4	21,0	60,4	22	67	60,4	21	60,4
5	8	4	29%	47%	24%	64	20	57,9	19	66	64,7	19,9	58,2
								Average	20	67	64	20	59
								Standard Deviation	3,08	3,22	4,55	2,33	4,49

Manufacturing of such component using Double-Double's is much simpler. The different bays will not have massive discontinuities at the bay edges to fulfill the symmetry conditions. Stacking the Double-Double plies could be done either from inside-out starting at the neutral axis or starting completely from inside or outside of the tool. Smooth, flush, step-free surfaces on one of the two sides of the laminate are thus made possible this way. See Fig. 7.

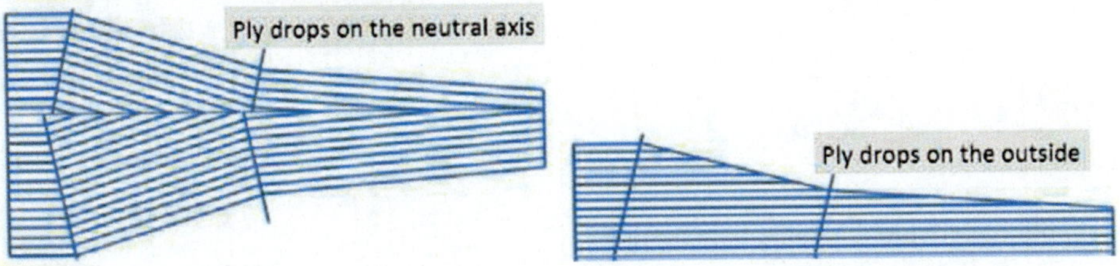

Fig. 7. Ply drop strategies using homogenized laminates

5.4 Optimization by *Lamsearch*

Traditional optimization strategies require a lot of repeated finite element simulation runs to evaluate samples with parameter changes. See Fig. 8 left. If surrogate models are used for optimization, many samples need to be created to derive those models as well, e.g. to develop response surfaces. A 10-ply sub-laminate of quad-axial style offers more than 25.000 (!) possible permutations for its stacking even for those 4 fixed ply angles. One reason for many developers not touching the legacy quad directions. Optimization this way thus remains a challenge, at least very time consuming.

The field equations with only two variable Double-Double angles and the trace-normalized stiffnesses enable a much simpler approach to find the optimum angles of a laminate with respect to strength, e.g. applying the Tsai-Wu failure criterion or last ply failure using the *Unit Circle Criterion*. This can be done by a "*Lamsearch*" algorithm.

At first for multiple loads on a single patch of laminate the best Double-Double angles can be found by simple sampling with fixed increments of the angles and "best of selection" with respect to strength in strain space. Multiple loads might either represent different locations on a structure up to the total of all finite elements involved in the optimization or using multiple load steps at just a single location.

After getting the optimum angles, the remaining tasks require linear scaling only – of local thickness to reach the required strength or selection of the best suited material to further reduce local thicknesses or scaling of the applied loads to reach failure. This step does not require additional finite element analyses.

This *Lamsearch* could be done with for samples with larger increments at first and later redone in a localized optimal domain by refining the angular increments. Any accuracy is possible. This process requires only one finite element analysis up front. Since only algebraic equations need to be solved for the remaining steps, *Lamsearch* requires only little computational effort.

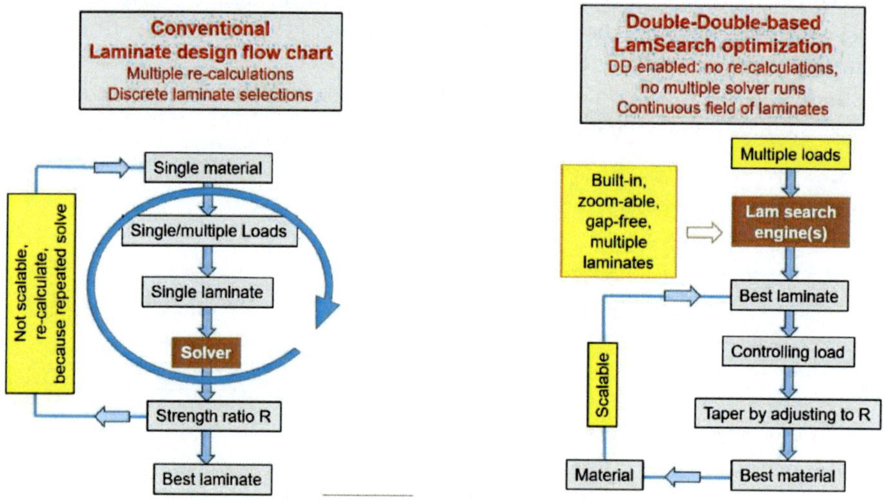

Fig. 8. Traditional optimization procedure of quad-axial laminates (left) and search-based optimization using Double-Double (right)

5.5 Manufacturing of double and double-double fabrics and tapes

Certainly, manufacturing of tapes with arbitrary fiber direction need to be provided for Double-Double. Karl Mayer GmbH from Germany is a manufacturer of such machines of non-crimp fabrics. Very small angles >15° are technically possible with reasonable tape width. Exactly 0° in tape direction is always possible as can be seen in Fig 9 left for three different tape widths (green: easily possible; orange: not recommended, but possible and leading to a high machine length; red: not possible because machine gets too long). Narrower tapes as semi-finishes products should be produced by slitting either as dry non-crimp fabrics of prepreg tapes. Chomarat in France already offers such fabrics as "C-Ply" using thin-plies, so lowest areal weights of the fabrics are achieved.

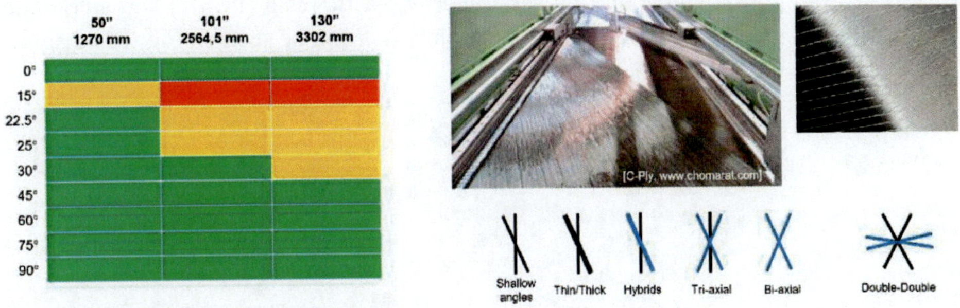

Fig. 9. Manufacturing of spread tow non-crimp fabrics (dry or prepreg) for Double or Double-Double semi-finished products

Layup of Double-Double tapes in one step or single Double tapes by using a two directional layup is straightforward. For thicker laminates this process is just repeated. Due to homogenization no complex layup with critical transitions at the bay boundaries is required. Layup is highly simplified. Tapering is much easier to achieve. Internal notches at bay boundaries are minimized. Such manufacturing process could be done using either prepregs or dry fabrics for infusion-based manufacturing. This layup process is well suited for automatic tape layering ATL.

6 Skin-grid structures

Skin-grid type panels have been in focus to achieve high performance lightweight structures for a long time. Massive milling of aluminum is still used for such airplane structures. Examples are found by the carry through box of A-330neo, isogrid structures used by McDonnelDouglas and NASA for Delta, Titan and Atlas space launchers and Skylab. Manufacturing using CFRP could not yet be demonstrated with sufficient accuracy and in a highly automatic fashion. Except for wounded vessels as reported from Russia using isogrids we are not aware of successful solutions for such skin-grid structures made of endless fiber composites. The later also has shown large material pileup and too much resin content at the grid intersections and between.

Double-Double might be used for stiffened panels as well. The grid can be interpreted as one of the doubles having fibers in two distinct directions in each layer. The problem to manufacture such structures is to avoid material pileup at the grid intersections. To avoid pile up, every second tape of a grid is discontinuous. This discontinuity reduces strength at the junctions to some content but keeps the stiffness almost unchanged. This is alike at every bay transition of laminated panels. The superior strength of such structures and its huge lightweight potential maximizes the advantage using Double-Double for high performance panel structures.

Fig. 10 (left) shows the first skin-grid panel as manufactured by MTorres, Pamplona, Spain as presented at the 2019 Paris air show. Tests at NASA have shown very high strength for this high-performance lightweight design [4]. The accuracy of the automatically manufactured skin and grid can be seen in the cuts at very left in Fig. 10.

This manufacturing in the meantime was also tried out for the first time at the Montreal based Concordia University by using thermoplastic resin (PEEK) and automatic tape placement to create a first grid intersection which can be seen on Fig. 10 at right.

Fig. 10. First automatically manufacture thermoset skin-grid panel from MTorres (left) and first thermoplastic CFRP at right (CF-PEEK)

References

1. Tsai SW, Melo JDD. An invariant-based theory of composites. Compos Sci Technol 2014;100:237–43.
2. Tsai SW, Melo JDD. Composite Materials Design and Testing - Unlocking mystery wit invariants. Stanford: Composites Design Group, Department of Aeronautics and Astronautics, Stanford University; 2015.
3. Tsai SW, Melo JDD, Sihn S, Arteiro A, Rainsberger R. Composite Laminates: Theory and practice of analysis, design and automated layup. Stanford: Composites Design Group, Department of Aeronautics and Astronautics, Stanford University; 2017.
4. Tsai SW et al. (2019) Composite double-double and grid/skin structures; compositesdesign.stanford.edu
5. Rother K, Tsai, SW, Seneviratne W. Manufacturing issues of structures based on double-double composite concepts. Presentation held at 4th ACM, Montreal 25.-26.4.2019
6. Arteiro A. et al. (2020) A case for Tsai's Modulus, an invariant-based approach to stiffness. Comp Struct 252 112683
7. Neuhäusler J, Rother K. Application of Tsai's Theory for the Laminate Optimization of an Aerospace Wing Box. Proceedings of the 2020 Munich Lightweight Symposium, Springer
8. NASA Report 4730 (April 1997): Global Cost and Weight Evaluation of Fuselage Side Panel Design Concepts; prepared by Boeing Commercial Airplane Group, Renton

Application of Tsai's Theory for the Laminate Optimization of an Aerospace Wing Box

Josef Neuhäusler[0000-0001-5205-3241] Klemens Rother[0000-0002-9643-4967]

Munich University of Applied Sciences, 80335 Munich
josef.neuhaeusler@hm.edu

Abstract. This work includes the application and evaluation of new methods to describe laminate stiffness and strength. Tsai et al. have shown that the trace of laminate stiffness is a rotationally invariant quantity and accurately describes the stiffness potential of a material [1]. In combination with a rotationally invariant strength criterion, the Unit Circle criterion, this allows a simple approach for the dimensioning of fiber composite structures [2]. The use of two biaxial, so called "double-double" [±ϕ/±ψ] sublaminates with the possibility of asymmetrical stacking sequences but homogenization of laminates further simplifies the design and manufacturing process of such structures [3].

The principles mentioned above are applied as an example for the optimization of an aerospace wing box. They are compared with classical optimization algorithms, an Evolutionary Algorithm (EA) and the Adaptive Response Surface Method (ARSM) [4]. The wing box is optimized with respect to stiffness while simultaneously minimizing weight. It is shown that the use of [±ϕ/±ψ] sublaminates instead of the traditional [0/±45/90] sublaminates can lead to a weight reduction of the composite skins of more than 10%. The simplified search algorithm based on the principles of Tsai et al. yields a different sublaminate than the classical optimization methods EA and ARSM. The computational effort though can be significantly reduced with the former.

Keywords: laminate optimization, trace-based sizing

1 Introduction

Tsai and Melo [5] showed that the mathematical trace of the stiffness matrix of an unidirectional ply $tr\mathbf{Q}$ is invariant to coordinate transformation. The same holds true for the traces of thickness normalized in-plane and flexural laminate stiffnesses: $tr\mathbf{Q} = tr\mathbf{A}^* = tr\mathbf{D}^*$. Furtherly they discovered that different Carbon Fiber Reinforced Plastics (CFRP) share very similar trace-normalized engineering constants $E_1^* = \frac{E_1}{tr\mathbf{Q}}$, $E_2^* = \frac{E_2}{tr\mathbf{Q}}$, $G_{12}^* = \frac{G_{12}}{tr\mathbf{Q}}$. That was the idea of the formation of a so called universal or master ply. For a list of several common materials, the trace-normalized longitudinal stiffness

S. Pfingstl et al. (eds.), *Proceedings of the Munich Symposium on Lightweight Design 2020*, https://doi.org/10.1007/978-3-662-63143-0_11

component Q_{11}^* had a variation compared to the mean value of 1.5% while the transverse and shear components appeared to show larger coefficients of variation of up to 16.4% [1].

Many investigations explained the significance of the knowledge of trace concerning novel laminate concepts [3], aircraft certification principles [2], optimization [6], strength prediction of notched specimen [7] or fracture mechanics [8].

Since the fundamental discovery of trace has been a benefit for so many applications in the design of composite materials researches have proposed to name such traces as an engineering constant, a material constant called Tsai's Modulus [1].

Besides the existence of an invariant description of stiffness – Tsai's Modulus – an invariant description of strength has been proposed [9] where the inner envelope of last ply failure (LPF) envelopes is used to generate a 1-parameter or 2-parameter Unit Circle (UC) criterion, see Fig. 1.

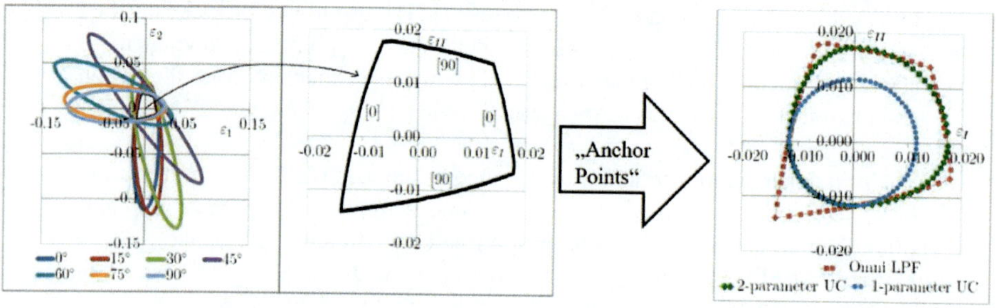

Fig. 1. Generation of the Unit Circle (UC) criterion from Tsai-Wu last ply failure (LPF) envelopes, T700/2510

A trace-based direct sizing approach has been presented in [10]. The authors incorporated trace and the UC criterion. With this approach, shortly explained in Fig. 2a), a baseline sublaminate with an initial thickness has to be selected. Further on the stiffness is derived for the baseline laminate using universal laminate factors from the universal ply. A linear elastic finite element analysis (FEA) is performed for each load case. For each finite element a maximum failure index can be determined. Finally, the initial uniform thickness profile can be scaled linearly, in case of in-plane loading, in order to meet a fail-safe design requirement. That final design may be scaled to any other baseline material using trace or to meet higher stiffness or strength requirements.

The next design idea is the use of double-double sublaminates $[\pm\phi/\pm\psi]$ and asymmetric stacking sequences. The advantages in terms of manufacturing, design and optimization have been extensively discussed in literature [2] [3] [6]. One big advantage of double-double sublaminates is that their stiffness components can be described analytically in closed form for arbitrary ply angles, see Fig. 2b). That is conceptually different than traditional carpet plot design for quasi-isotropic $[0/\pm45/90]$ sublaminates. That opens a whole new idea for laminate design and optimization. Multiple FEA's can be avoided by employing stiffness transformations within the continuous field of $[\pm\phi/\pm\psi]$ sublaminates instead of multiple FEA's within the discrete design alternatives for the traditional sublaminates.

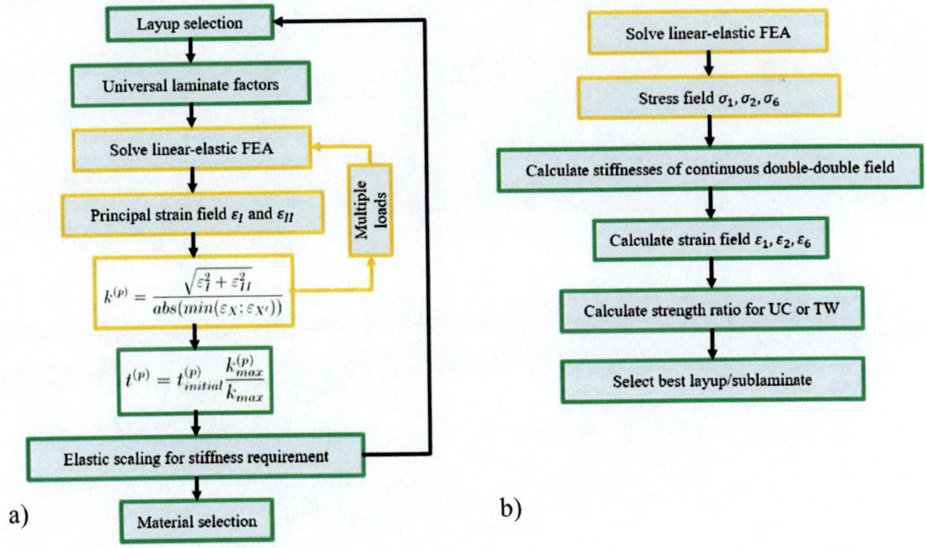

Fig. 2. Algorithms used for laminate design of a wing box: a) trace-based direct sizing algorithm, b) laminate search algorithm

2 Optimization of a Wing Box

The design case in this study was taken from a diploma thesis [11] at the Munich University of Applied Sciences Munich. In that thesis, two construction methods of a carbon fiber composite wing box for a fighter plane were compared. The first type of construction consisted of a sandwich core, two skins with a $[0/\pm45/90]$ sublaminate and a front and rear stringer with rivets. The second type is a pure stringer stiffened wing box without a sandwich core. In both cases, an optimization was carried out using FEA and numerical optimization procedures in order to minimize weights and reserve factors of the constructions. The sandwich construction appeared to outperform the stringer stiffened version.

The primary goal in this section is to find a better double-double sublaminate $[\pm\phi/\pm\psi]$ than the traditional quasiisotropic $[0/\pm45/90]$ sublaminate in terms of a lower weight.

2.1 Finite Element Model

The sandwich construction is remodeled according to the technical drawings in the appendix of [11]. It is then transferred to Abaqus[1] where the preprocessing, solution and postprocessing are done.

[1] Abaqus is a trademark of Dassault Systemes

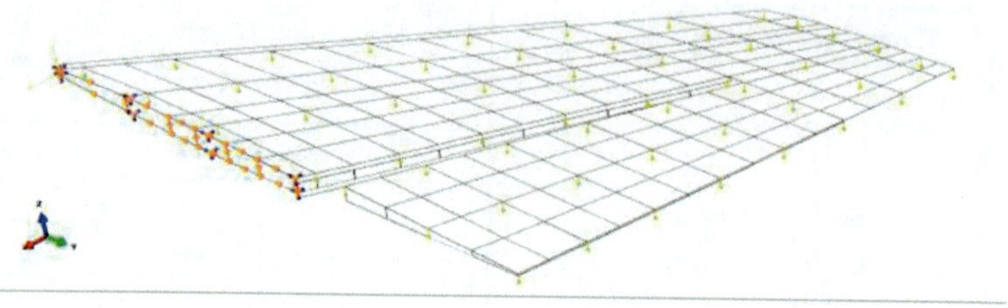

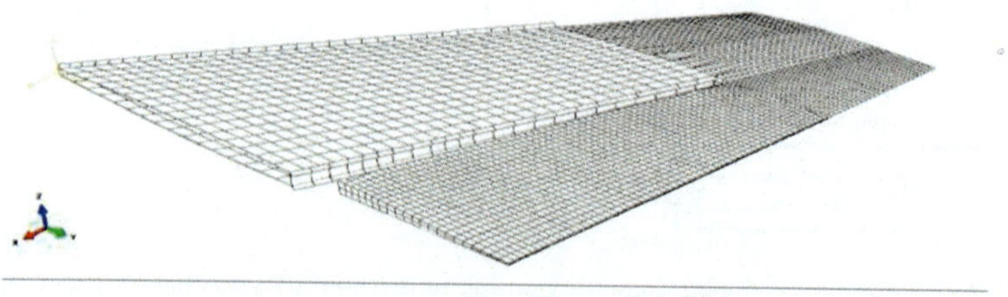

Fig. 3. a) Geometry, loading and boundary conditions, b) Finite element mesh

The finite element model (FE model) consists of the following parts with their respective Abaqus finite element types [12]:

- a glass fiber/phenolic resin honeycomb core type HRP, discretized with quadratic tet C3D10 elements
- two carbon/epoxy skins, discretized with quadratic S8R elements and bonded with a tie constraint to the honeycomb core
- one front and one rear aluminum stringer, discretized with quadratic S8R elements and bonded with a tie constraint to the honeycomb core and the skins
- an aluminum wing box tip and rudder, discretized with quadratic S8R elements and bonded with a tie constraint to the honeycomb core and the skins

Fig. 3a) depicts the geometry and loading/boundary conditions where the yellow arrows represent a pressure distribution along the wing in the form of concentrated nodal forces. Two load cases (LC's) are imposed:

- LC1: Subsonic with shove
- LC2: Supersonic with rudder movement and shove

For both load cases the boundary condition is a fully fixed support where the wing box typically is attached to the fuselage.

The finite element mesh can be seen in Fig. 3b) where the upper and lower composite skins are meshed with an element seed size of 80 mm. This element size has been verified with a preliminary study on the convergence of the global stress profile.

The material data for the FE model besides aluminum is given in Table 1. Composite section properties with the material data for ACA are assigned to the lower and upper skin in order to make a ply-by-ply analysis possible. The HRP data is applied to the solid core elements using engineering constants in Abaqus. Honeycomb cores typically have different shear moduli in different orientations. Similar to the original thesis, the lower shear modulus in W-direction is assigned to the whole core.

Table 1. Material data of ACA CFRP and HRP honeycomb

ACA	E_x	E_y	E_s	ν_{xy}	tr	t_{Ply}	ρ
	[MPa]	[MPa]	[MPa]	[−]	[MPa]	[mm]	$[\frac{t}{mm^3}]$
	140000	6000	1300	0.3	149000	0.0625	1.55E − 9
	X	X'	Y	Y'	S		
	[MPa]	[MPa]	[MPa]	[MPa]	[MPa]		
	1300	1000	25	100	40		
HRP	E_c^0	E_s^0	ρ				
	[MPa]	[MPa]	$[\frac{t}{mm^3}]$				
	311.5	34.5	6.40E − 11				

Note that the mathematically correct tensor value for the trace of the ACA material is given in Table 1. In case we want to calculate trace by using a master ply value, the value of trace would be:

$$tr_{calc} = \frac{E_x^0}{0.885} = \frac{140000}{0.885} = 158000 \quad (1)$$

That value differs more than 5% from the trace calculated with classical laminate theory.

Note that the FE-model does not exactly represent the FE-model from the original thesis regarding the size of the finite elements. Loads and dimensions are all measured by hand according to the available technical drawings. However, the model is an adequate foundation for the investigations on the different laminate optimization algorithms.

2.2 Laminate Optimization with EA and ARSM

The first optimization of the wing box is carried out with two rather conventional programming algorithms. The two algorithms employed in this work are the Adaptive Response Surface Method (ARSM) and an Evolutionary Algorithm (EA).

ARSM is basically an adaptive design of experiment by starting with an initial design point and a global sampling scheme around the starting point. Then polynomial regression models approximate the model response and the new designs are sampled around the supposed optimum. The algorithm converges as the parameter space of the following samplings shrinks further [4].

EA can be best described by a “survival of the fittest” strategy. Mutation, recombination and selection lead to the evolution of the algorithm and eventually to an optimum design [4].

A key for the performance in terms of required design calculations to find a global optimum is a reasonable start design. Such a start design point might be the global minimum estimated with an initial sensitivity analysis. Both methods are covered in the software optiSLang[2] which is used for the laminate optimization of the wing box. The default settings of optiSLang were used. More detailed information about the software implementation and optimization methods can be found in [4].

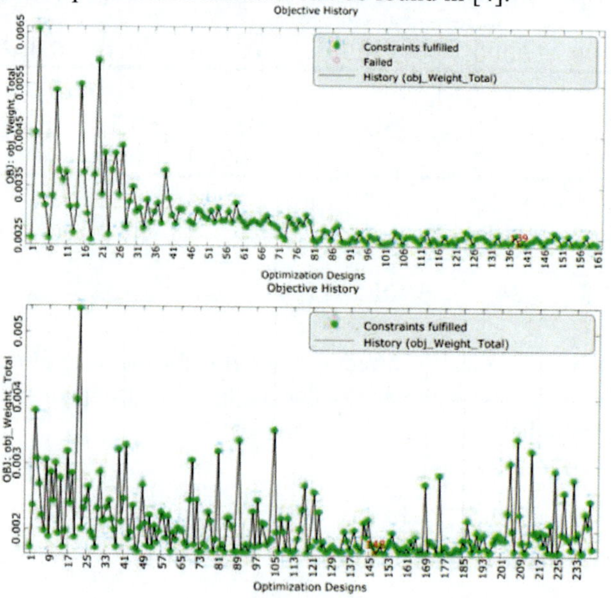

a)

b)

Fig. 4. a) History plot for the Adaptive Response Surface Method ARSM, b) History plot for the Evolutionary Algorithm EA

The optimization algorithms implemented in the software optiSLang directly take over the FEA solving process. Two design variables are defined: both double angles $\pm\phi$ and $\pm\psi$. The model response is the calculated weight which is determined by the trace-based sizing algorithm in Fig. 2a). The goal is to minimize the weight of the skins. A further restriction is that the ply angles can only vary in 7.5° steps between 0° and 90° for a better comparability with the laminate search optimization in the following section. A reasonable quasi-isotropic double-double sublaminate $[\pm22.5/\pm67.5]$ was chosen as start design.

The history plot of the ARSM in Fig. 4a) illustrates that the global optimum was found after 139 iterations. Contrary to that the EA in Fig. 4b) required 148 iterations. Note that both algorithms continue with further iterations after the minimum response

[2] optiSLang is a trademark of Dynardo GmbH, Weimar, Germany owned by ANSYS, Inc., Canonsburg, USA

has been found. That behavior is typical for direct optimization methods [13]. ARSM and EA lead to the same best double-double sublaminate:

$$DD_{opt,ARSM} = DD_{opt,EA} = [\pm 22.5/\pm 60]$$

That sublaminate is close to the quasi-isotropic $[\pm 22.5/\pm 67.5]$ which has the same in-plane stiffness as the traditional $[0/\pm 45/90]$ sublaminate from the original thesis.

2.3 Laminate Optimization with Laminate Search

For the second optimization approach the laminate search algorithm as displayed in Fig. 2b) is used. In contrast to the optimization with ARSM and EA only one FEA is required to find the optimum sublaminate. The upper and lower skins are now meshed with homogenous shell elements without shell section properties in order to obtain the global stress field $\sigma_{1,2,6}$. These stresses are output in the global coordinate system.

Fig. 5 shows the resulting R values also known as strength ratios for different baseline laminates with the Unit Circle strength criterion for last ply failure (LPF). The strength ratios on the z-axis are plotted over different combinations of $[\pm\phi/\pm\psi]$. The angles ϕ and ψ are given in 7.5° angle increments for illustration purposes. 0 represents the double angle $[\pm 0]$ and 13 represents the double angle $[\pm 90]$. In fact, the strength ratio could also be plotted as a continuous surface for all ply angles since the stiffnesses and consequently the strength ratios can be expressed analytically in a closed form.

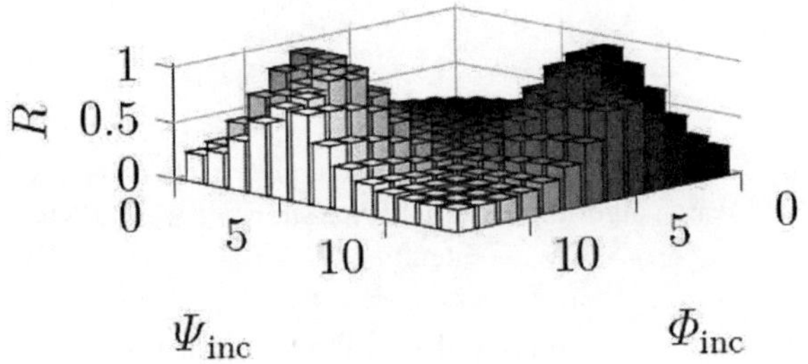

Fig. 5. Controlling R values plotted over angle increments, 7.5° increments from 0 (0°) to 13 (90°) using Unit Circle (UC) strength criterion

The R values or strength ratios could also be described as safety factors. Knowing this helps to select the best sublaminate by choosing the combination of $\pm\phi$ and $\pm\psi$ with the highest safety factor. The best double-double sublaminate using the laminate search (LS) and Unit Circle criterion appears to be:

$$DD_{opt,LS,UC} = [\pm 30/\pm 60]$$

The sublaminate $DD_{opt,LS,UC} = [\pm 30/\pm 60]$ differs from the optimum sublaminate found by ARSM and EA $DD_{opt,ARSM} = DD_{opt,EA} = [\pm 22.5/\pm 60]$. A finer resolution in Fig. 5 in terms of smaller angle increments would help to find a sublaminate which is closer to the ARSM and EA. Compared to the ARSM (139 FEA's) and EA (148

FEA's) only one FEA is required using the laminate search algorithm which decreases the computational effort significantly.

2.4 Thickness Scaling

After a suitable sublaminate has been found, the initial profile with a uniform thickness value can be scaled to the final thickness profile by using UC, as described in Fig. 2a). The final scaled thickness profiles for sublaminate $DD_{opt,LS,UC} = [\pm 30/\pm 60]$ and sublaminate $DD_{opt,ARSM} = DD_{opt,EA} = [\pm 22.5/\pm 60]$ are given in Fig. 6 where each finite element is assigned with a uniform thickness. An increased thickness along the rear aluminum stringer and towards the fixed support, where the wing box was attached to the fuselage, can be observed.

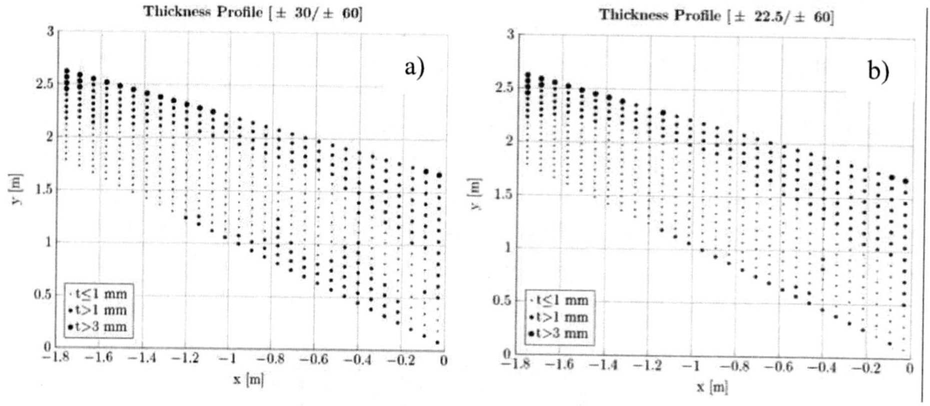

Fig. 6. Thickness profiles of the skins with manufacturable thicknesses after the scaling procedure with, a) laminate search algorithm with 1 FE-iteration and b) NLP algorithm with 139 FE-iterations

The thickness profiles in Fig. 6 are fail-safe thickness profiles with respect to the UC strength criterion for LPF. Table 2 shows the weights of the fail-safe thickness profile. The lightest design alternative results from the optimization with ARSM and EA with 4.4 kg/skin. Next would be the quasi-isotropic sublaminate with 4.6 kg/skin and finally the sublaminate from the laminate search algorithm with 4.8 kg/skin.

Table 2. Weights of the scaled skin profiles

Sublaminate	Weight per skin
$DD_{opt,ARSM} = [\pm 22.5/\pm 60]$	4.4 kg/skin
$DD_{opt,EA} = [\pm 22.5/\pm 60]$	4.4 kg/skin
$DD_{Quasi-iso.} = [\pm 22.5/\pm 67.5]$	4.6 kg/skin
$DD_{opt,LS,UC} = [\pm 30/\pm 60]$	4.8 kg/skin

The sublaminate from ARSM and EA was used for further investigations on the buckling resistance. Eventually a linear buckling analysis was performed and the thicknesses had to be increased in order to prevent the structure from buckling, see Fig. 7.

Note that the focus of this work was the minimization with respect to the UC strength criterion. Recently a paper was published presenting multi-objective pareto optimization incorporating further requirements for strength, buckling resistance and wing-tip deflection under multiple load cases [6].

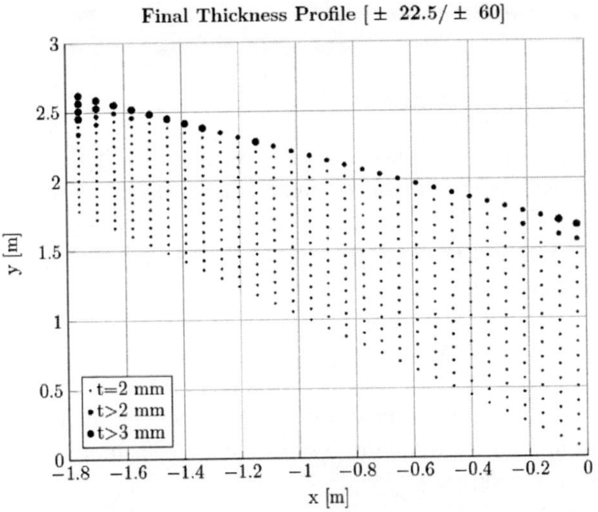

Fig. 7. Final thickness profile of the skins

Eventually the final fail-safe and buckling resistant thickness profile in Fig. 7 can be compared with the original thesis, see Table 3. A weight reduction of 13.9% of the composite skins has been achieved in the current work.

Table 3. Skin weights of the original thesis and the current work

Thesis/Sublaminate	Weight per skin
Original Thesis/[0/±45/90]	8.6 kg/skin
Current Work/[±22.5/±60]	7.4 kg/skin

3 Discussion of the Wing Box Optimization

3.1 Material Selection

At the end of the sizing procedure material selection can be performed which is enabled by the trace-based direct sizing approach. Whereas in traditional laminate optimization the composite designer has to choose the material before the optimization procedure.

For an exemplary new material with a trace 20% lower than the original ACA material, the thickness profile has to be increased by 20%. One always has to keep in mind that direct linear scaling is only possible in case of mainly in-plane loading conditions. In case of pure bending, scaling is no more linear but related with a cubic root function.

In our investigated benchmark case we only focused on the mainly in-plane loaded skins and therefore thickness scaling was assumed to be linear.

Another advantage for selecting the material at the end of the optimization procedure is material ranking. Since the finally optimized solution may be transferred to any other material via trace, the best performing material in terms of e.g. manufacturability, material cost, layup speed could be selected.

3.2 Comparison with Aluminum

A direct comparison of the composite skins with equally stiff aluminum skins can only be done if a quasi-isotropic sublaminate such as the double-double laminate $DD_{Quasi-iso.} = [\pm 22.5/\pm 67.5]$ is assumed. Since we furtherly assume linear scaling the weight of an equally stiff aluminum skin using the normalized Young's modulus of isotropic materials of 0.337 [14]:

$$W_{alu} = \frac{W_C tr_C E_{1C}}{26.6 \rho_C} = 5.5\ \text{kg} \tag{2}$$

with tr_c and E_{1c} being stiffness measures of the composite in GPa and ρ_c the density of the composite in g/cm^3. Aluminum weighs almost 20% more than the weight of the equally stiff composite skin.

3.3 Laminate Homogenization

One limiting factor regarding asymmetrical stacking sequences so far certainly is homogenization. A major argument for using mid-symmetric laminate architectures is that the coupling matrix $[B]$ is exactly zero in theory. That is the reason why mid-symmetric laminates are thermomechanically stable and not affected by warping or spring-in, especially during the curing process.

Asymmetric laminates become quasi-homogenous when the number of plies is increased until the thickness normalized coupling matrix $[B^*]$ approaches zero. However, the availability of thin NCF with 50 g/sm makes homogenization possible much faster than with traditional thick plies with e.g. 150 g/sm. A double-double $[\pm\phi/\pm\psi]$ sublaminate will homogenize faster than a traditional six, eight or ten plies thick $[0_p/\pm 45_q/90_r]$ sublaminate.

Warping due to thermomechanically induced residual stresses is actually caused by many factors. Firstly, dimensions of the structure and laminate affect thermomechanical distortion [15]. The length to thickness ratio of a laminated plate highly dictates the warping or snap through of asymmetric laminates. Hyer [16] derived estimations of deflections for thin unsymmetric laminates based on data fitting of the classical laminate theory. The predictions however did not conform with the experimental data. A publication by Jun and Hong [15] solved the coefficients for the differential equation of the deflection by minimizing the potential energy involving effects of thermal expansion with a Newton-Raphson iteration. They observed that the shapes of thin cross-ply laminates flatten with increasing number of plies.

A second group of influences on warping are process parameters. Controlled cure cycles are essential to avoid early gelation. Different coefficients of thermal expansion between part and tool may also affect thermal warping [17].

York [18] presented a group of sublaminates that are intrinsically symmetric and have $[B^*] = 0$ but at least eight plies. It is clear that a fast homogenization in terms of repeated ply stacking of asymmetric laminates is only possible with sublaminates having a small number of plies, such as double or double-double sublaminates.

A recently published paper [6] studied thermal warpage of double and double-double laminates compared to traditional $[0_p/\pm 45_q/90_r]$ laminates. The authors concluded that double and double-double sublaminates outperform traditional laminates in terms of thermal warpage in the investigated cases of a wing-box.

Conclusion

The trace-based direct sizing algorithm has been successfully applied to the weight minimization of a wing box. The weight of the composite skins has been reduced by using $[\pm\phi/\pm\psi]$ sublaminates compared to the traditional quasi-isotropic sublaminates. Programming algorithms such as EA and ARSM lead to a lower weight of the composite skins than the laminate search algorithm in the investigated design case. However, the computational effort can be reduced significantly with the latter. Asymmetric stacking sequences require minimum thicknesses in order to achieve thickness homogenization of a structure in order to reduce thermomechanically induced residual stresses. Required thicknesses for homogenized laminates with asymmetric stacking sequences need further clarification.

Declaration of Competing Interest

The authors declare that they have no known competing financial interests or personal relationships that could have appeared to influence the work reported in this paper.

References

1. A. Arteiro and N. Sharma et al., "A case for Tsai's Modulus, an invariant-based approach to stiffness," *Composite Structures,* vol. 252, 2020.
2. S. W. Tsai et al., Composite Double-Double and Grid/Skin Structures: Low Weight/Low Cost Design and Manufacturing: A Stanford/Mtorres Invention, Stanford: Composites Design Group, 2019.
3. S. W. Tsai and J. D. D. Melo, Composite Materials. Design and Testing: Unlocking mystery with invariants, Stanford: Composites Design Group, 2015.
4. Dynardo GmbH, Weimar, Method for multi-disciplinary optimization and robustness analysis, Weimar, 2020.

5. S. W. Tsai and J. D. D. Melo, "An invariant-based theory of composites," *Composites Science and Technology,* vol. 100, pp. 237-243, 2014.
6. S. Shrivastava, N. Sharma, S. W. Tsai and P. M. Mohite, "D and DD-drop layup optimization of aircraft wing panels under multi-load case design environment," *Composite Structures,* vol. 248, 2020.
7. C. Furtado, A. Arteiro, M. A. Bessa, B. L. Wardle and P. P. Camanho, "Prediction of size effects in open-hole laminates using only the Young's modulus, the strength and the R-curve of the 0° ply," *Composites Part A: Applied Science and Manufacturing,* vol. 101, pp. 306-317, 2017.
8. D. Dalli, G. Catalanotti, L. F. Varandas, B. G. Falzon and S. Foster, "Compressive intralaminar fracture toughness and residual strength of 2D woven carbon fibre reinforced composites: New developments on using the size effect method," *Theoretical and Applied Fracture Mechanics,* vol. 106, 2020.
9. S. W. Tsai and J. D. D. Melo, "A unit circle failure criterion for carbon fiber reinforced polymer composites," *Composites Science and Technology,* vol. 123, pp. 71-78, 2016.
10. S. W. Tsai, A. Arteiro and J. D. D. Melo, "A trace-based approach to design for manufacturing of composite laminates," *Journal of Reinforced Plastics and Composites ,* vol. 35, pp. 589-600, 2016.
11. K. Rother, Bauweisenvergleich eines Seitenleitwerkkastens aus Kohlefaserverbundwerkstoff für ein Jagdflugzeug. Engl.: Comparison of the construction methods of a carbon fibre composite fin box for a fighter plane, Munich: Munich University of Applied Sciences, 1985.
12. "Abaqus/CAE User's Manual. 6.11," Dassault Systemes, 2011.
13. P. W. Christensen and A. Klarbring, "An Introduction to Structural Optimization," *Solid mechanics and its applications,* vol. 153, 2009.
14. S. W. Tsai, S. Sihn and J. D. D. Melo, "Trace-based stiffness for a universal design of carbon-fiber reinforced composite structures," *Composites Science and Technology,* no. 118, pp. 23-30, 2015.
15. W. J. Jun and C. S. Hong, "Residual stresses and warping in unsymmetric laminates with arbitrary lay-up angles," *Applied Stress Analysis,* 1990.
16. M. W. Hyer, "Some Observations on the Cured Shape of Thin Unsymmetric Laminates," *Journal of Composite Materials,* vol. 15, pp. 175-194, 1981.
17. C. Albert and G. Gernlund, "Spring-in and warpage of angled composite laminates," *Composites Science and Technology,* vol. 62, pp. 1895-1912, 2002.
18. C. B. York, "On tapered warp-free laminates with single-ply terminations," *Composites Part A: Applied Science and Manufacturing,* vol. 72, pp. 127-138, 2015.

Author Index

S. Pfingstl et al. (eds.), *Proceedings of the Munich Symposium on Lightweight Design 2020*, https://doi.org/10.1007/978-3-662-63143-0